10 LED PROJECTS
FOR GEEKS

10 LED PROJECTS
FOR GEEKS

BUILD LIGHT-UP COSTUMES, SCI-FI GADGETS, AND OTHER CLEVER INVENTIONS

EDITED BY JOHN BAICHTAL

**no starch
press**

SAN FRANCISCO

Printed in USA

First printing

22 21 20 19 18   1 2 3 4 5 6 7 8 9

ISBN-10: 1-59327-825-X
ISBN-13: 978-1-59327-825-0

Publisher: William Pollock
Production Editor: Riley Hoffman
Interior Design: Beth Middleworth
Developmental Editor: Liz Chadwick
Technical Reviewer: Steven Bingler
Copyeditor: Rachel Monaghan
Compositor: Riley Hoffman
Proofreader: James Fraleigh

The following images are reproduced with permission:
Figure 1 by afrank99 (*https://commons.wikimedia.org/wiki/File:Verschiedene_LEDs.jpg*; CC-BY-SA-2.0);
Figure 4 by Phillip Burgess (*https://learn.adafruit.com/assets/30740*, *https://learn.adafruit.com/assets/30733*, and *https://learn.adafruit.com/assets/30727*; CC-BY-SA-3.0);
Figure 5 by Phillip Burgess (*https://learn.adafruit.com/assets/30715*; CC BY-SA-3.0);
Figure 6 by SparkFun (*https://www.sparkfun.com/datasheets/Components/LED/COM-09590-YSL-R531R3D-D2.pdf*);
Figure 8 by Peter Halasz (*https://commons.wikimedia.org/wiki/File:Seven_segment_02_Pengo.jpg*; CC-BY-SA-3.0);
Figure 9 by BigRiz (*https://commons.wikimedia.org/wiki/File:Fibreoptic.jpg*; CC-BY-SA-3.0);
Figure 10 by Jack Lee (*https://commons.wikimedia.org/wiki/File:Red_and_green_traffic_signals,_Stamford_Road, _Singapore_-_20111210.jpg*; CC-BY-SA-3.0);
Figure 11 by Led-neolight (*https://commons.wikimedia.org/wiki/File:Br20_1.jpg*; CC-BY-SA-3.0);
Figure 12 by the Raspberry Pi Foundation (*https://www.raspberrypi.org/learning/images/components/raspberry-pi.png*; CC-BY-SA);
Figures on page 21 by Adafruit Industries (*https://cdn-shop.adafruit.com/1200x900/2377-02.jpg*; *https://cdn-shop .adafruit.com/1200x900/284-03.jpg*);
Figure 2-1, Figure 2-2, and Figure 4-1 by Adafruit Industries;
Figure 6-2 by Thomas Wydra (*https://commons.wikimedia.org/wiki/file:ir_led.jpg*; CC-BY-SA-3.0);
Figure on page 139 by MithrandirMage (*https://en.wikipedia.org/wiki/Viewing_frustum#/media/File:ViewFrustum.svg*; CC-BY-SA-3.0).
Circuit diagrams and schematics were created using Fritzing (*http://fritzing.org/*).

For information on distribution, translations, or bulk sales, please contact No Starch Press, Inc. directly:

No Starch Press, Inc.
245 8th Street, San Francisco, CA 94103
phone: 1.415.863.9900; info@nostarch.com
www.nostarch.com

Library of Congress Cataloging-in-Publication Data
A catalog record of this book is available from the Library of Congress.

TO MY CO-AUTHORS, FOR THEIR
CLEVERNESS AND PATIENCE. -JB

ABOUT THE CONTRIBUTORS

JOHN BAICHTAL has written or edited over a dozen books, including the award-winning *The Cult of LEGO* (No Starch Press, 2011), LEGO hacker bible *Make: LEGO and Arduino Projects* (Maker Media, 2012) with Adam Wolf and Matthew Beckler, as well as *Robot Builder* (Que, 2013) and *Hacking Your LEGO MINDSTORMS EV3 Kit* (Que, 2015). John lives in Minneapolis with his wife and three children.

KAAS BAICHTAL became interested in computers and electronics around age 12, taking every available electronics class. As a technician Kaas worked mostly in the entertainment industry, doing equipment repairs and travelling system installs for theatrical dimmer manufacturers AVAB America and Electronic Theatre Controls (ETC) and multimedia integrator BBI Engineering. Kaas has run her own servers at home since 1998 and specializes in writing custom code to solve real-life problems.

MATTHEW BECKLER is a computer engineer who lives in Minneapolis with his wife and two cats. His day job usually consists of writing firmware for fancy microcontrollers, and he is a co-founder and engineer at a fun side-hustle called Wayne and Layne, where he and Adam Wolf design and sell electronic kits and help create interactive museum and art exhibits. He holds a BS in computer engineering from the University of Minnesota and an MS and PhD in electrical and computer engineering from Carnegie Mellon University.

KRISTINA DURIVAGE is an independent software developer by day and a hardware hacker by night, specializing in data visualization and making the world a brighter place with LEDs. Her work is collected at *http://portfolio.gelicia.com/* and her opinions and cat pictures can be found on Twitter, @gelicia.

LENORE M. EDMAN is a co-founder of Evil Mad Scientist Laboratories, a family-run company that designs, produces, and sells hobby electronics kits, drawing machines, and retrotechnological objects. She writes for the accompanying project blog on the topics of electronics, crafts, cooking, science, robotics, and anything else that catches her fancy. Many of the blog's projects have been featured at science and art museums and in *Make*, *Wired*, and *Popular Science* magazines. She holds a BA in interdisciplinary studies (English and Greek).

MIKE HORD has been working at SparkFun Electronics designing products and projects for makers for several years. His making skills run the gamut from metalworking, woodworking, and 3D printing to coding and circuit design. When not creating his next Big Hack, he's raising two small children to question the veracity of everything except the need for toothbrushing.

JAMES FLOYD KELLY is a full-time technology writer in Atlanta, Georgia. He has written more than 25 books on a mix of topics that includes open source software, LEGO robotics, basic electronics, Arduino programming, and more. He and his wife have two young boys who are showing the early signs of maker-ness.

MICHAEL KRUMPUS has a master's degree in computer science and 25 years of experience as a software engineer. He discovered a passion for electronics design later in life and formed a small electronics company, nootropic design, where he designs and manufactures innovative electronics for hobbyists, designers, educators, and industry. Michael is based in Minneapolis.

WINDELL H. OSKAY is the co-founder of Evil Mad Scientist Laboratories, a Silicon Valley company that has designed and produced specialized electronics and robotics kits since 2007. Evil Mad Scientist Laboratories also runs a popular DIY project blog, and many of its projects have been featured at science and art museums and in *Make*, *Wired*, and *Popular Science* magazines. Windell served as a founding board member of OSHWA, the Open Source Hardware Association. Previously, Windell has worked as a hardware design engineer at Stanford Research Systems and as a research physicist in the Time and Frequency Division of the National Institute of Standards and Technology. He holds a BA in physics and mathematics from Lake Forest College and a PhD in physics from the University of Texas at Austin.

ADAM WOLF is a co-founder of and engineer at Wayne and Layne, where he designs DIY electronics kits and interactive exhibits. He also does computer engineering and embedded systems work at an engineering design services firm in Minneapolis. When he isn't making things blink or helping computers talk to each other, he's spending time with his wife and sons.

ABOUT THE TECHNICAL REVIEWER

STEVEN BINGLER is a software engineer with a focus on low-level and embedded systems and has always had an interest in small, fun projects. He earned his master's degree from the University of Florida and in his spare time enjoys bicycling, tinkering, and finding new places to eat.

CONTENTS

CONTENTS IN DETAIL

INTRODUCTION
BY MATTHEW BECKLER

Welcome to *10 LED Projects for Geeks*, featuring 10 projects of varying complexity that you can build using electronics. You'll be using the Raspberry Pi and Arduino to create all kinds of gadgets and inventions in this book. In this introduction you'll learn how to get started with both boards, including installing the necessary software, and you'll meet the simple electronic component that is the focus of this book: the LED.

At the end of this introduction, you'll also find a list of popular suppliers for parts you'll need in the projects.

ALL ABOUT LEDS

Before you start building cool projects using LEDs, let's take a moment to define them. An *LED* is a circuit component that converts electricity into light. LEDs come in all colors, shapes, and sizes, and are useful in a variety of projects. Most LEDs are used for *indication*, meaning they indicate the status of something to a nearby human, such as "door is open," "radio is on," or "security system is armed." Nowadays, larger and brighter LEDs are available that can provide *illumination* of an object or region. Figure 1 shows some basic LEDs used for indication.

FIGURE 1:

A variety of single LEDs

LED stands for *light-emitting diode*. There are a few words here to unpack, but let's start with the word *diode*. In a circuit, a diode works like a one-way valve, permitting electricity to flow in only one predefined direction. If you try to send electricity in the other direction, it won't flow. Diodes convert alternating current (AC) electricity into direct current (DC), a process called *rectification*. Diodes were originally designed for use in AM radio receivers to rectify radio frequency signals, but are now used in many different applications where electrical current must flow in only one direction. For example, diodes in many consumer electronics help protect the circuitry against batteries inserted backward. Most of the time, however, the diode aspect of an LED is not important. The rest of the name, *light-emitting*, simply means that this type of diode is designed to light up when electricity passes through.

How LEDs Work

An LED is a *semiconductor device*, built with crystalline metals like silicon and germanium, plus small amounts of impurities that change the characteristics of the material. These impurities are added to the semiconductor in a process called *doping*. The added element, called the *dopant*, changes the base material into either a *p-type* or *n-type* semiconductor.

The distinction between p-type and n-type depends on whether the dopant adds excess free electrons, or causes a deficit of free electrons. The place where these two types of doped semiconductor meet is called the *P-N junction*, and this is where light is created. Electrical energy passing through the junction excites some of the charge carriers, which emit light when they return to their nonexcited state in a process called *electroluminescence*.

The exact chemical composition of the semiconductor material determines the band gap of the materials involved, which in turn determines the color of light produced. The first LEDs, developed in the 1960s, emitted light in the infrared end of the color spectrum, which is invisible to the human eye but still useful for scientific experiments. Later developments of other semiconductors and dopants introduced new, visible colors, resulting in mass production of red LEDs in the late 1960s and yellow LEDs in the early 1970s.

A high-brightness blue LED was very difficult to develop, and was considered the Holy Grail of the LED world until 1994, when three Japanese scientists finally figured out the secret to a blue LED. They received the 2014 Nobel Prize in physics for this accomplishment!

A BRIEF HISTORY OF ARTIFICIAL LIGHTING

Throughout human history, the development of artificial lighting has been incredibly important in the advancement of technology. Without artificial lighting, it's almost impossible to do much productive work after sunset, whether that involves crafting, entertainment, or education. For a long time, the only way to produce artificial light was to burn some sort of fuel, such as wood, beeswax, oil, coal gas, or kerosene. This changed with the development of electricity distribution networks in the late 1800s when electric lighting became widespread.

Early electric lighting consisted of *incandescent light bulbs*, which produce light by heating a thin filament wire until it glows white-hot, typically in a bulb filled with inert gas (to extend the filament life). These bulbs convert only a small percentage of the input energy into light, wasting the rest as heat. Incandescent light bulbs eventually gave way to *gas-discharge bulbs*, including fluorescent lights, neon signs, and high-intensity discharge (HID) lamps such as streetlights, which pass an electric arc through a tube of sealed gas to produce light more efficiently. Gas-discharge bulbs typically require a high voltage, so they are inconvenient for portable and low-power applications. With a *light-emitting diode*, nothing gets white-hot and no gas arcs are involved, just highly efficient light production.

LED Configurations

LEDs are very popular for all types of artificial light, and come in many configurations. Let's explore some of the most common types you'll encounter.

Single and Bicolor LEDs

The basic *single LED* configuration has two connections: the anode and cathode. If sufficient voltage is applied, the LED lights up. Pretty simple! Single LEDs come in all different shapes, sizes, and colors, and some, such as the second and fourth ones shown in Figure 1, have extra plastic to help direct the generated light through an opening in the side of an electronics case.

A *bicolor LED* contains two LEDs in the same two-lead package, typically with the individual LEDs connected anode-to-cathode, as shown in Figure 2.

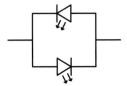

Usually the two LEDs have different colors, enabling the circuit to display two different colors using only two leads. For example, if current travels through the circuit from left to right, the lower LED turns on. If current travels through the circuit from right to left, the upper LED turns on.

Another fun type is the *candle flicker LED*, which contains a small computer chip that turns the LED on and off in a flickering pattern, designed to emulate the natural flicker of a candle. You've probably seen these LEDs at fancy restaurants or wedding receptions, or anywhere you want to have the ambiance of a candle without the risk of fire.

Red-Green-Blue LED

The *red-green-blue (RGB) LED* packages three LEDs into one, and actually allows you to create a whole spectrum of colors, not just red, green, and blue. For example, you can combine red and green light to make yellow, green and blue light to create cyan, and red and blue light to make violet. Mixing all three colors of light at the same intensity produces white light.

By combining these three colors of light at varying levels of brightness for each individual color, you can make millions of colors; this is the same method your TV and phone use to display full-color images. Usually, to vary the brightness, each color is switched on and off very rapidly, which tricks our eyes into seeing each color at a different level of intensity. The simplest RGB LED has four connections, one for each of the three colors plus one connection common to all three LEDs. This common connection can be either a *common anode (CA)* or *common cathode (CC)*, which describes the internal connections of a four-pin RGB LED. Figure 3 shows a diagram of the connections of a common anode RGB LED.

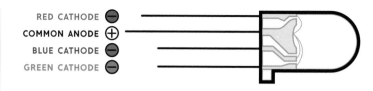

FIGURE 3:

A common anode
RGB LED contains red,
green, and blue LEDs
and has four wire leads

RED CATHODE
COMMON ANODE ⊕
BLUE CATHODE
GREEN CATHODE

Another common type of RGB LED is the *digitally controlled RGB LED*, which contains a tiny computer chip used to control the brightness of each color. Your microcontroller communicates with the chip in each RGB LED and sets the desired color. The chips in the RGB LEDs are often either the WS2812B or the SK6812, and software libraries exist for almost any system you want to use with them (including Arduino, Raspberry Pi, and many others).

The biggest benefit of using digitally controlled RGB LEDs is that you can "set and forget" them, so your microcontroller can update the LED with the latest desired color, then move on to other tasks instead of having to constantly update the LED brightness. You can also chain RGB LEDs together to form large strings or arrays of full-color RGB LEDs, making them useful for art projects and outdoor displays.

These LEDs are available in a variety of form factors, as shown in Figures 4 and 5.

FIGURE 4:

A variety of RGB LEDs:
tiny surface-mount
chips; four-pin through-
hole package; LEDs
preattached to small
circuit boards

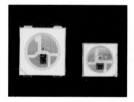
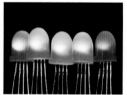

FIGURE 5:

A flexible grid of digitally
controlled RGB LEDs

These figures show the digitally controlled RGB LEDs in a variety of packages, including tiny surface-mount chips, four-pin through-hole packages, LEDs attached to little circuit boards, and even LEDs preattached in flexible grids.

A final type of RGB LED is a two-connection, internally controlled LED. These are a good option if you're looking for really cheap RGB LEDs with only two leads. These LEDs have a small control chip inside, but unlike with the digitally controlled RGB LEDs, you can't actually control what they do. For example, as soon as you provide power to these LEDs, the internal control chip starts displaying a prerecorded sequence of colors on the RGB LED, usually cycling through each primary and secondary color with a nice color fade between each one.

Reading an LED Datasheet

The manufacturer of any LED usually provides a technical datasheet full of facts and figures. It can seem overwhelming, but don't worry! For the purposes of this book, you need to know only two important numbers to use the LED safely. Figure 6 shows a typical LED datasheet that you'll use to find the critical details, which are highlighted here in red.

Absolute Maximum Ratings: (Ta=25℃)

ITEMS	Symbol	Absolute Maximum Rating	Unit
Forward Current	I_F	20	mA
Peak Forward Current	I_{FP}	30	mA
Suggestion Using Current	I_{SU}	16-18	mA
Reverse Voltage (V$_R$=5V)	I_R	10	uA
Power Dissipation	P_D	105	mW
Operation Temperature	T_{OPR}	-40 ~ 85	℃
Storage Temperature	T_{STG}	-40 ~ 100	℃
Lead Soldering Temperature	T_{SOL}	Max. 260℃ for 3 Sec. Max. (3mm from the base of the expoxy bulb)	

Absolute Maximum Ratings: (Ta=25℃)

ITEMS	Symbol	Test condition	Min.	Typ.	Max.	Unit
Forward Voltage	V_F	I_F=20mA	1.8	---	2.2	V
Wavelength (nm) or TC(k)	$\Delta \lambda$	I_F=20mA	620	---	625	nm
*Luminous intensity	I_V	I_F=20mA	150	---	200	mcd
50% Viewing Angle	$2\theta 1/2$	I_F=20mA	40	---	60	deg

FIGURE 6:

An LED datasheet with forward current and forward voltage highlighted in red

An LED produces light when electrical current flows through it, but too much current can overheat and damage it. Using the *forward voltage* and the *maximum current* parameters from the datasheet, we can figure out what size resistor we need to operate the LED safely.

Forward Voltage

The *forward voltage*, or V_f, of an LED is the minimum amount of voltage the LED needs to start conducting electricity. It measures the difference in voltage between the anode and cathode when it's on. If you apply a voltage below V_f, the LED will not conduct any electricity or light up. When the LED is operating normally, the voltage across the LED will be equal to V_f. We'll use this fact later on when we start calculating circuit parameters.

The V_f of an LED depends on the chemistry of the materials used, which also determines the color of the LED. For example, red LEDs have a typical V_f of 1.5 to 2 V, while blue or white LEDs tend to have higher V_f values in the range of 2 to 3 V.

Maximum Current

If your voltage exceeds V_f, the LED conducts as much electricity as possible, which can overheat and damage it. To prevent this, we need a second value from the datasheet: the *maximum current*, which is the maximum electrical current that can pass through the LED without overheating it.

Most LEDs have a *maximum steady current* and a higher *maximum pulsed current*. Usually, we use the maximum steady current for our calculations. We measure current in units of amperes, or *amps* (abbreviated as *A*), and in thousandths of an amp, or *milliamps* (abbreviated as *mA*). Current is the amount of electricity that is flowing through a device, and it determines the brightness of an LED. For example, most small LEDs light up with only 10 to 30 mA of current. Some larger LEDs and LED assemblies need hundreds of milliamps or even several amps of current!

Resistance, LEDs, and Ohm's Law

To safely power an LED, we need to provide at least V_f voltage across the LED but also limit the current to ensure it remains below the datasheet's max current allowed. The easiest way to do this is to add a resistor "in series" (inline) with the LED. A *resistor* is the simplest possible circuit component, with a very simple relationship between the amount of voltage across the resistor, the current flowing through the resistor, and the amount of resistance of the resistor. This relationship is called *Ohm's law* and can be expressed by the following equation:

$$\text{voltage} = \text{current} \times \text{resistance}$$

We can simplify this expression using the letter *I* to represent current, giving us the familiar $V = I \times R$ equation for Ohm's law. Resistance is measured in units called *ohms*, abbreviated as Ω.

Let's use a practical example to understand this relationship. Imagine a resistor that has 1,000 Ω of resistance. If we apply 12 V to this resistor, how much current will flow through it? To answer this question, we simply substitute the variables in the equation with the information given:

$$\text{voltage} = \text{current} \times \text{resistance}$$

$$\text{current} = \frac{\text{voltage}}{\text{resistance}}$$

$$\text{current} = \frac{12\ \text{V}}{1{,}000\ \Omega}$$

$$= 0.012\ \text{A}$$

$$= 12\ \text{mA}$$

Using Ohm's law, you can see that 12 mA of current will flow through a resistor with 1,000 Ω of resistance when we apply 12 V. Let's try this same calculation again, but with 36 V this time:

$$\text{current} = \frac{36\ \text{V}}{1{,}000\ \Omega}$$

$$= 0.036\ \text{A}$$

$$= 36\ \text{mA}$$

Now, let's see how resistors can help us protect an LED from drawing too much current.

Calculating Resistance to Use with an LED

If we pair an LED with a resistor, they work together to limit the current through both devices. Figure 7 shows how this works.

FIGURE 7:

Using a resistor in
series (inline) with an
LED to limit the current
and protect the LED

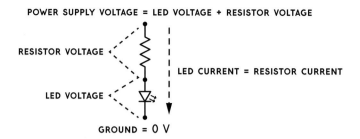

POWER SUPPLY VOLTAGE = LED VOLTAGE + RESISTOR VOLTAGE

RESISTOR VOLTAGE

LED CURRENT = RESISTOR CURRENT

LED VOLTAGE

GROUND = 0 V

Notice that the current through the LED is the same as the current through the resistor. You can see that the overall power supply voltage is also equal to the voltage across the LED plus the voltage across the resistor. We can write these two facts as two simple equations to describe our circuit:

$$\text{LED current} = \text{resistor current}$$

$$\text{power supply voltage} = \text{LED voltage} + \text{resistor voltage}$$

We already know that the LED current needs to be below the maximum it can handle. For this circuit, let's say we want 20 mA of LED current maximum. We also know the LED's forward voltage (V_f) from the datasheet. For this example, let's say the forward voltage is 2 V, and the power supply voltage is 5 V. With these facts, we can fill in some variables in the two equations.

$$20 \text{ mA} = \text{resistor current}$$

$$5 \text{ V} = 2 \text{ V} + \text{resistor voltage}$$

$$\text{resistor voltage} = 5 \text{ V} - 2 \text{ V}$$

So we have computed that the resistor current is 20 mA, and the resistor voltage is 3 V. All we need to do now is to calculate what resistance value we need for the resistor, using Ohm's law.

$$\text{resistor voltage} = \text{resistor current} \times \text{resistor resistance}$$

$$3 \text{ V} = 20 \text{ mA} \times \text{resistor resistance}$$

Rearranging, we find:

$$\text{resistor resistance} = \frac{3 \text{ V}}{0.020 \text{ A}}$$

$$\text{resistor resistance} = 150 \text{ }\Omega$$

This tells us we need 150 Ω of resistance to keep the LED current at 20 mA with a 5 V power supply. Keep in mind that most resistors are accurate only to within 5 percent of their specified value. This means that a 1,000 Ω resistor value will have an actual resistance between 950 and 1,050 Ω, and some resistors might even have a 10 percent tolerance for the accuracy of their value. If your calculations result in a resistance value that isn't commonly available, round up to the nearest available value. Rounding up the resistance may slightly reduce the current, but it's better than slightly increasing the current and possibly overloading the LED!

We can generalize this calculation for a situation where you have an LED with inline resistor connected to a power supply:

$$\text{resistance} = \frac{\text{power supply voltage} - \text{LED forward voltage}}{\text{LED maximum current}}$$

That's a lot of math. But how does this technology get used? Let's look at some of the ways people use LEDs.

LEDs in the World

We use LEDs in a wide variety of applications in both obvious and not-so-obvious ways. For example, many digital clock displays on appliances like the microwave or radio have a *seven-segment display* for each digit of the time. This display consists of seven LEDs arranged in the shape of a blocky numeral 8, as shown in Figure 8. By turning each of the seven segments on or off, we can display all the digits from 0 through 9.

FIGURE 8:

A seven-segment display consists of multiple LEDs integrated into a single plastic package, enabling the display of digits 0 through 9.

A less obvious LED application is their use in many *fiber optic cables* and *infrared remote controls* for communications. The vast undersea internet cables that connect cities and continents consist of thin glass strands with an LED (or sometimes a laser) at each end, transmitting pulses of light through the cable to communicate data. Figure 9 shows an example of fiber optic cables.

FIGURE 9:

A bundle of fiber optic cables

Infrared (IR) remote controls for televisions and other appliances use a special type of LED that creates invisible infrared light that only the appliance's sensor can detect. Because many video cameras are sensitive to infrared light, we use IR LEDs in security cameras to generate nighttime illumination that remains invisible to human eyes.

Many new traffic signals and automobile lights now use LEDs, as shown in Figure 10, which helps to increase visibility and reduce both power consumption and waste heat generation.

FIGURE 10:

An LED traffic light

With the relatively recent development of pleasing "warm white" LEDs, shown in Figure 11, many households and businesses are starting to adopt LED lighting. These LED bulbs can be over 10 times more efficient than an incandescent light bulb, and their reduced energy use makes up for their higher purchase price within just a few years of use.

FIGURE 11:
A white LED light bulb designed to fit in a standard lamp socket

Now that you've learned about the different types of LED and how they work, you're almost ready to dive in to the projects! In addition to a wide variety of LEDs, some of the projects in this book will need a Raspberry Pi or Arduino. These next sections guide you through setting up both. If you already have these set up, or want to set them up later when you need them, you can skip these sections for now and go straight to the first project!

GETTING STARTED WITH THE RASPBERRY PI

This section will show you how to set up your Raspberry Pi with a formatted SD card and the Raspbian operating system.

What You'll Need

Raspberry Pi is perhaps the most well known of the small-form-factor computers on the market right now. About the size of a credit card, the Raspberry Pi is a tiny computer that includes a processor, USB and other connectors, and input/output pins, as shown in Figure 12.

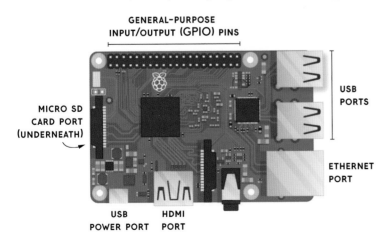

GENERAL-PURPOSE INPUT/OUTPUT (GPIO) PINS

USB PORTS

MICRO SD CARD PORT (UNDERNEATH)

ETHERNET PORT

USB POWER PORT

HDMI PORT

FIGURE 12:
A Raspberry Pi single-board computer

There are a few different models of Pi. Any Pi will do for the purposes of this book.

Like any computer, the Pi needs a keyboard, mouse, display, power supply, and some storage for the operating system. You can use any USB mouse and keyboard as well as any HDMI display, like a computer monitor or a TV. The power supply is a micro USB connector like you'll find in many cell phones, so you can reuse an old phone charger if you already have one. While many computers have a spinning magnetic hard disk drive for storage, the Pi needs a separate micro SD card to store the operating system (OS) and files you create. You'll therefore need to get an 8GB or 16GB micro SD card—the bigger, the better. We'll be following along with the official Raspberry Pi Software Guide (*https://www.raspberrypi.org/learning/software-guide/quickstart/*).

Installing Raspbian with NOOBS

The most popular OS for Raspberry Pi is called *Raspbian*, and you can easily install it by copying files onto the micro SD card from another computer. The easiest way to get started with Raspberry Pi is by using the *NOOBS* ("New Out-of-the-Box Software") system, which lets you choose which OS to install.

To install NOOBS, visit the Raspberry Pi website's Downloads section (*https://www.raspberrypi.org/downloads/noobs/*). Then select the **NOOBS (Offline and network install)** option. Don't select NOOBS LITE, which doesn't have Raspbian included by default. To download NOOBS, click the **Download ZIP** option. You can select **Download Torrent** instead if you're comfortable using BitTorrent software (this helps reduce bandwidth costs for the Raspberry Pi Foundation). Either way, when the download finishes, open the ZIP archive using your unzip program and extract the files to your computer. Put them somewhere easy to find, like on your Desktop or in your *Documents* directory.

Preparing the SD Card

Next, we need to format the micro SD card to prepare it to receive the NOOBS files. Even if your SD card is brand new, I recommend formatting it, as it needs to be entirely empty to receive the OS. If your computer runs Windows or macOS, download SD Formatter 4.0 from the SD Association website (*https://www.sdcard.org/*) by going to the Downloads section and finding the formatter for your current

NOTE

If your computer runs Linux, you can use any disk formatting utility to format the SD card using a Master Boot Record (MBR) and a single FAT partition.

OS. Follow the instructions to install the software and use it to format your micro SD card.

After formatting the micro SD card, drag and drop the extracted NOOBS files onto the micro SD card drive. Wait until the files finish copying, then safely remove the micro SD card and insert it into the Raspberry Pi.

Starting Up Your Pi

If you haven't already, plug in your keyboard and mouse. Connect to your display with the HDMI cable. If you have an Ethernet cable available to connect your Pi to the internet, plug it in at this time. Finally, plug in the micro USB power cable, and the Raspberry Pi should start up. Unlike most computers, the Raspberry Pi doesn't have a power button, but automatically powers up when you plug in the power connector. Refer to Figure 12 to see which ports are which.

If this is the first time you're using this micro SD card, the NOOBS system will ask you to select which operating system you want to use. Because it's designed especially for the Raspberry Pi and built in to the NOOBS system, Raspbian is the OS I recommend, but there are other options that may require an internet connection to download.

Now that you're done setting up your Raspberry Pi is and the Raspbian OS, you're ready to build a bunch of projects. Project 6 in this book uses the Pi. After that, there are a ton of cool projects documented online you can try, both on the official Raspberry Pi website (*https://www.raspberrypi.org/resources/*) and on other websites like Instructables and Hack-a-Day.

GETTING STARTED WITH THE ARDUINO AND THE ARDUINO IDE

Some projects in this book use the Arduino as their brains. The Arduino is less complicated than the Raspberry Pi in that it doesn't have the computer and networking capabilities that the Pi has, but simply runs one program over and over until you unplug it. However, the Arduino excels at many hardware projects, especially when high-precision timing is required (such as talking to digital RGB LEDs). Let's explore what Arduino is, cover how to install it, and get familiar with the Arduino workflow.

What Is Arduino?

The Arduino project began in 2003 with the aim of making it easy for people to get started with electronics, programming, and interactive design. The project created a variety of open source microcontroller circuit boards, along with a series of software libraries tied together in an easy-to-use *integrated development environment (IDE)*. Arduino boards provide an interface between the digital world and the real world, giving users a simple way to read data from sensors like buttons, switches, and thermometers, as well as drive actuators such as lights, motors, and LED displays. An entire ecosystem has sprung up around the core Arduino system, offering a huge set of add-on hardware and software you can use to design the next great robot, art installation, or musical instrument.

The basic Arduino workflow goes something like this. First, write the programming code, called a *sketch*, in the Arduino IDE program on your computer. Second, connect the Arduino circuit board to your computer using a USB cable. Third, select the **Upload** button in the Arduino IDE to transfer the code into the Arduino board. The code will then automatically start running, reading sensors and driving actuators to interact with the real world.

The most popular Arduino board is the *Arduino Uno*, which we'll use as an example in this section to illustrate the process of uploading code into the board. Arduino boards have a main processor chip (such as the Atmel ATMega328 on the Arduino Uno) as well as a USB chip that handles the interface between the computer and the main processor. These boards can be connected to your computer with just a USB cable.

Other Arduino-compatible boards, such as the EMSL Diavolino and the SparkFun Arduino Pro Mini, don't include the USB chip, usually to reduce cost. For these boards, you need a separate piece of hardware to provide the USB interface between the computer and the main processor chip. Because the most popular USB interface chip for the Arduino is made by a company called FTDI, we call this USB interface an *FTDI cable*, or an *FTDI Friend*.

Installing and Using the Arduino IDE

To get started with Arduino, you first need to install the Arduino IDE from the official website (*https://www.arduino.cc/*) under the Software section. I recommend you download the offline IDE rather than relying on the web editor. Packages are available for Windows, macOS, and Linux distributions, including Raspberry Pi.

Figure 13 shows the Arduino IDE running on Windows 10.

The interface should look similar across different platforms. The important toolbar buttons are as follows:

❶ **checkmark** Checks for any errors and verifies that the sketch can be compiled

❷ **right arrow** Uploads and transfers the sketch into the circuit board

❸ **magnifying glass** Opens the serial monitor to see information sent by the Arduino board

The programming code for the sketch goes into the large text area in the center of the screen. The programming language used for Arduino is called *Wiring*, which is similar to the C++ language.

What an Arduino Sketch Looks Like

The Arduino IDE comes with a large number of example sketches, intended to demonstrate all the different things that an Arduino can do. The simplest sketch, called *Blink*, blinks an LED light. Most Arduino boards have a built-in LED that you can turn on and off using the sketch code, and the Blink sketch uses this built-in LED as an output signal. This sketch is a great way to test out a new Arduino board and confirm that the IDE is properly set up.

To open the Blink example sketch, use the menus to select **File ▸ Examples ▸ 01.Basics ▸ Blink**, as shown in Figure 14.

FIGURE 14:

Selecting and opening
the Blink sketch

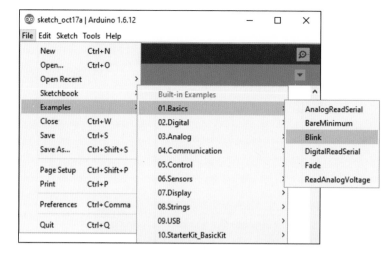

Every Arduino sketch has at least two sections of programming code, which we call *functions*. The first required function is setup(), which we call only once when the board is first powered up or reset. The setup() function is where we place all activity having to do with initialization and configuration, such as initializing the functionality of each pin and configuring the features of an external library. The other required function is loop(), which is called repeatedly after setup() has been called once. You can create your own functions, too, to help organize your sketch code. Listing 1 shows the complete program listing for the Blink sketch.

LISTING 1:

The complete code
listing for the Blink
sketch

```
// the setup function runs once when you press reset
// or power the board
void setup() {
  // initialize digital pin LED_BUILTIN as an output
  pinMode(LED_BUILTIN, OUTPUT);
}

// the loop function runs over and over again forever
void loop() {
  digitalWrite(LED_BUILTIN, HIGH);   // turn the LED on (HIGH
                                     // is the voltage level)
  delay(1000);                       // wait for a second
  digitalWrite(LED_BUILTIN, LOW);    // turn the LED off by
                                     // making the voltage LOW
  delay(1000);                       // wait for a second
}
```

Here, you can see that the setup() function configures the LED's pin as an output pin that can turn the LED either on or off. This is in contrast to an input pin, which would read in a HIGH, LOW, or even a varying voltage applied to it.

The loop() function contains two other Arduino functions. The simplest is delay(), which pauses the program execution for the specified number of milliseconds (thousandths of a second). Each time the program reaches the delay(1000) line, it stops and waits for 1 second before continuing.

The digitalWrite() function works only on a previously configured output pin, and it sets the output voltage to either HIGH or LOW. The LOW voltage is always 0 V (usually called *ground*), and the HIGH voltage depends on the specific circuit board, but is usually 5 V or 3.3 V.

In sum, this sketch first configures the LED's pin as a digital output, and then turns the LED on, waits for 1 second, turns the LED off, waits 1 second, and repeats the process over and over again until the Arduino board is reset or the power is disconnected.

Configuring Your Board and Port

Before we compile and upload the sketch into the Arduino board, we need to use the Tools menu to tell the Arduino IDE what kind of Arduino board we're using. Figure 15 shows how you can select the Arduino Uno by choosing **Tools ▸ Board**. Here, we're using the Uno as an example, but be sure to select the entry that matches the board you're using.

NOTE

Lines and phrases beginning with // are comments and do not affect the code's execution.

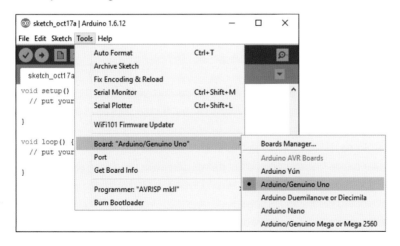

FIGURE 15:

Using the Tools menu to select which Arduino board is connected

Some Arduino boards have additional options you can use to specify the amount of available memory, how fast it should run, or which processor type is installed, but the Uno doesn't have any options like this. All boards do have the Port option, which specifies which USB port the device is connected to. This is platform dependent, of course, because ports have different names on macOS, Windows, and Linux, but most modern Arduino boards will be able to transmit their name, which should show up in the Port menu as shown in Figure 16.

FIGURE 16:

Selecting the correct
Arduino board from
the list of connected
devices in the
Port menu

After you configure the board and port, the final step is to use the Upload button to prepare (*compile*) and upload the sketch code into the attached Arduino board. To do this, press the right-arrow button on the toolbar, or select the **Sketch ▸ Upload** menu option. You should be able to see the compiler output in the black area below the sketch code, which also displays how much space remains for program storage and global variables. Because each Arduino board allots different amount of space for storing the compiled sketch and the global variables, be sure to pay attention to these numbers to keep track of how full the chip is getting.

Although the Blink sketch doesn't transmit any debugging messages to the serial monitor, many other sketches use the serial monitor as a way to provide feedback on the running program. Some sketches even take input from the serial monitor and use it to influence how the program runs. In this way, you can make your sketches interact with the computer or any other device that can talk to a USB serial port.

Now that you've learned the basic workflow of using the Arduino IDE to compile and upload code, let's touch on a couple of add-ons used in this book.

Some of the projects in this book use Arduino add-ons for particular tasks. For example, the Adafruit Trinket is used for Project 2: The Desktop UFO, and the Adafruit Lilypad, which is often used for developing wearables, appears in Project 9: Wearable Timing Bracer. Each project that uses an add-on will explain how to set it up and use it.

ARDUINO BOARDS WITHOUT A BUILT-IN USB CONNECTION

As mentioned, some Arduino-compatible boards skip the USB interface chip to save cost or reduce power consumption, or both. One popular board like this is the *Arduino Pro Mini*, created by SparkFun. The Arduino Pro Mini (shown here; left) requires a separate board, the FTDI Friend (shown here; right) to provide the USB interface with the computer. These two boards work together to provide a low-cost Arduino board with a removable (and reusable) USB interface.

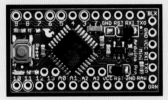 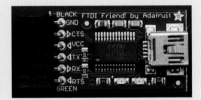

The photos are aligned as the boards would be when connected together. Simply solder in some header pins on the right side of the Arduino Pro Mini so that they can connect to the socket on the FTDI Friend. Use the Arduino IDE's Boards menu to select **Arduino Pro or Pro Mini**. This should enable the Processor option just below the Board option; be sure to pick the correct processor type and speed.

RECOMMENDED SUPPLIERS

Throughout this book, each project lists a small number of online stores as a source for the components used in it. Still, every store has products unique to it, so more sources are better! The following are online stores serving worldwide or regional markets.

United States

Adafruit (New York; *www.adafruit.com*) This electronics store has a ton of LED products and an impressive array of microcontrollers, breakout boards, and kits.

BGMicro (Texas; *www.bgmicro.com*) We love the unusual and obsolete parts found in this quirky store.

Digi-Key (Minnesota; *www.digikey.com*)   Get lost in the vast array of parts found in this huge online store.

Evil Mad Scientist (California; *www.evilmadscientist.com*) Kitmakers and component sellers par excellence. Owners Lenore Edman and Windell H. Oskay also cowrote Project 1 of this book.

Jameco Electronics (California; *www.jameco.com*)   Around since 1974, Jameco might literally be your dad's electronics store. It has a great paper catalog.

Mouser Electronics (Texas; *www.mouser.com*)   If you want a rare sensor or a thousand of a component, check with this mega-distributor.

SparkFun Electronics (Colorado; *www.sparkfun.com*)   Founded by a college student, SparkFun has exploded into an online retailer of kits and individual components.

Tower Hobbies (Illinois; *www.towerhobbies.com*)   Champaign-based Tower Hobbies is a radio-control and drone superstore.

Europe

Elextra (Denmark; *www.elextra.dk*)   This fun store has kits and tools and a wider-than-normal selection of electronic products.

Farnell/Newark element 14 (United Kingdom; *www.newark.com*) Based out of Leeds, Farnell sells electronic components throughout the world. Chicago-based Newark is its sister company.

Maplin (United Kingdom; *www.maplin.co.uk*)   Electronics retailer Maplin has over 200 stores in the UK and sells online as well.

Play Zone (Switzerland; *www.play-zone.ch*)   This hobby electronics site, available in English and German, has everything from building sets to Raspberry Pis to individual components.

RS Components (UK; *https://uk.rs-online.com/web*)   This electronics superstore boasts half a million products.

Asia

AliExpress (China; *www.aliexpress.com*)   There is no way to enumerate all the crazy things you can buy from eBay-like AliExpress.

DealExtreme (Hong Kong; *www.dx.com*)   Everything imaginable available and at eyebrow-raising prices—what could go wrong?

HobbyKing (Hong Kong; *www.hobbyking.com*)   This is a radio-control and drone superstore with warehouses in the US.

Seeed Studio (Shenzhen; *www.seeedstudio.com*)   Not just an online store, Seeed also runs business incubator Seeed Propogate as well as Seeed Fusion, a prototyping shop.

FOUR SIMPLE PROJECTS

BY LENORE EDMAN AND WINDELL OSKAY

IN THIS CHAPTER, YOU'LL CREATE FOUR SIMPLE LED PROJECTS.

LED-LIT SEA URCHINS

EDGE-LIT HOLIDAY CARDS

DARK-DETECTING LED

ELECTRIC ORIGAMI

There's a reason LED projects are among the earliest projects aspiring electronics hobbyists encounter: they offer simplicity as well as a glowing payoff. The following projects were featured on the Evil Mad Scientist Laboratories blog (*https://www.evilmadscientist.com/*). These simple projects will prepare you to tackle the more advanced projects found later in the book. First we'll show you how to make an LED-Lit Sea Urchin, and then you'll create an Edge-Lit Holiday Card, a Dark-Detecting LED, and Electric Origami.

#1: LED-LIT SEA URCHINS

At a beach shop during one vacation, we came across some tiny sea urchin shells, thin and light as eggshells. What to do with them? Light them up with LEDs, of course!

Seashell lamps aren't necessarily new, but they're usually made with large, heavy, and colorful shells. In contrast, these sea urchins are smaller and lighter, and illumination shows off their natural beauty.

Get the Parts

For this project, you'll need to find some small sea urchin shells, or some other similar container with walls thin enough to be translucent. The ideal size is about 4 inches across.

- LEDs (for example, Evil Mad Scientist *https://emsl.us/743* or SparkFun P/N 12062)

- Coin cell CR2032 batteries (SparkFun P/N 338)

- Small, light sea urchin shells

Each urchin has a hole in the bottom large enough to fit a pretty good size LED, although not necessarily the battery as well. When the shells are lit, they should look something like Figure 1-1.

FIGURE 1-1:
Different-colored LEDs make for an assortment of urchins.

Build It

First we'll connect our LEDs to watch batteries, a combination known as *throwies* because they're small and cheap enough that you could theoretically throw them without worry. We won't ask you to throw anything in this project, but you'll use the throwies to illuminate the urchin shell.

1. **Make the throwies.** Select an LED of any color. The longer leg is positive and the shorter leg is negative. Simply connect the leads to the battery terminals, taping the positive leg to the positive battery side and the negative leg to the negative side, and you have a tiny light, as shown in Figure 1-2. Bend the leads so that the LED points up, away from the battery.

FIGURE 1-2:
Combining a battery and an LED makes for quick and easy illumination.

2. **Assemble the urchins.** Insert the throwie in the hole at the bottom of the shell. If your battery doesn't fit, just poke the head of the LED through and rest the shell on the battery. Repeat this process for as many shells and colors as you like.

Voilà! The urchins are complete and glowing dramatically, showing off their elaborate shell structures. Your lamps should last about one to two weeks depending on the color of the LED.

#2: EDGE-LIT HOLIDAY CARDS

In this project, you'll make your own LED-lit holiday cards using clear plastic and a simple LED assembly. Figure 1-3 shows a few cards we created. The technique we employ, commonly called *edge-lit*, involves placing a lit LED on the edge of an etched piece of acrylic. This gives the etched area a nice glow without visible wires or other hardware.

FIGURE 1-3:

Create LED cards for
your loved ones.

Get the Parts

You'll need just a few supplies to make your LED holiday card.

- LEDs with clear lenses, preferably 3 mm or rectangular (for example, *https://emsl.us/533* or SparkFun P/N 08285)
- Clear plastic (approximately 2.5 × 3 inches, 1/16 inch thick)
- Electrical tape
- Coin cell CR2032 battery (SparkFun P/N 338)
- Pen
- Paper and cardstock
- Scissors
- Hobby knife

Build It

To create your own holiday card, first you need a design. You'll score this design into the plastic and the LED will light it from the bottom edge, reflecting off the score lines to create the glowing effect. You'll need to keep the design simple (like the Christmas tree and snow-flake designs in Figure 1-3), as scoring into the plastic isn't the most precise drawing method.

You'll begin with a piece of clear plastic sized at roughly 2.5 × 3 inches (65 × 75 mm) and 1/16 inch (1.6 mm) thick. We used acrylic, but you can also use polycarbonate, polypropylene, or a number of other clear plastics. You can cut a piece out of a thick-walled clear plastic container or get material like this at the hardware or hobby store. Our clear acrylic came with protective blue film on both sides.

1. **Cut a piece of plastic.** If your sheet of plastic is bigger than 2.5 × 3 inches, you'll need to cut it down. This particular material cannot be cut with scissors—it cracks and shatters—so you'll need to score it with a sharp hobby knife or box cutter and a metal straightedge, and then snap it quickly along the scored line. Using that method, it's pretty quick to cut an appropriately sized piece of plastic from a larger sheet. Figure 1-4 shows the plastic in its protective film.

FIGURE 1-4:
Cut a piece of plastic to match your card design.

2. **Create the paper template.** Trace the size of your plastic piece onto a sheet of paper. Then, draw a second rectangle inside the first (Figure 1-5), leaving a gap of about 0.5 inch (13 mm). Your drawing needs to fit well within this inner rectangle, to ensure that the final design is nicely centered on the plastic.

FIGURE 1-5:
Trace the piece of plastic, then draw a rectangle inside it.

3. **Draw the design and trace it.** When your paper drawing is finalized, place it on top of the plastic and trace it with the hobby knife.

    Obviously, how cool your shape looks will influence how cool the final product looks. Figure 1-6 is just a simple tree pattern—not too tricky. For the snowflake, you can print out and trace some real snowflake patterns.

FIGURE 1-6:

Draw your design inside the inner rectangle.

4. **Attach an LED and battery.** This project uses the same classic LED throwie arrangement shown in the LED-Lit Sea Urchins project. Attach an LED directly to the leads of a CR2032 lithium coin cell as described in "Build It" on page 27, but this time don't bend the leads at a right angle. Now, when you position the bright LED at the edge of the plastic (shown in Figure 1-7), it lights up the areas that you've scored.

5. **Tape up the edges.** Keep holding the LED to the bottom edge of the plastic so the light goes into the scored lines. Next, use black electrical tape to restrict light from going in other directions. Put down a piece of tape, sticky side up, and stick the plastic plate to the top half of the tape with the LED below it. Next, place a matching piece of tape on top, forming a tight seal around the LED to keep it pointed into the plastic (Figure 1-8).

    Next, seal the other three sides of the plastic plate with black tape the same way, as shown in Figure 1-9. Note that the tape only needs to cover the very edge of the plastic. After you've finished this process, the single LED will light up the design etched into the plastic without leaking much light elsewhere.

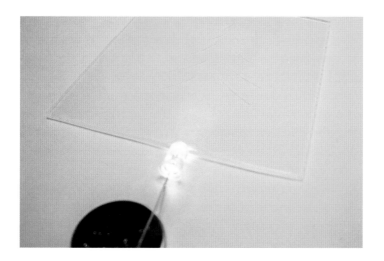

FIGURE 1-7:
Lighting the design from the bottom edge

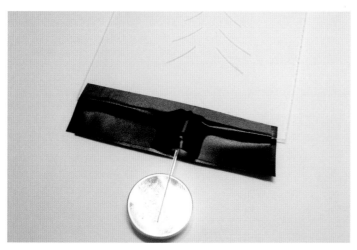

FIGURE 1-8:
Tape up the LED so no light escapes

FIGURE 1-9:
Tape up the edges of the design

6. **Prepare the paper card.** Next, you need to start the card itself. We used 9 × 12 inch heavy Bristol cardstock, but other heavy opaque papers will work as well, and the size isn't that critical. Fold the paper in half twice to make a classic greeting card shape with two layers on the front and back fold, each about 4.5 × 6 inches or so.

Place your completed edge-lit plastic assembly on top of the folded card to make sure it's going to fit, and determine where you're going to position it.

Next you'll cut a window in the front of the card so that the plastic design can show through. Cut out the inner rectangle from your template from step 2 and trace around it onto the front of your card. Unfold the card before you cut to make sure you cut only the front layer. Score the outline with the hobby knife and then carefully cut it out with either the hobby knife or scissors. You'll attach the edge-lit plastic assembly behind the window you just cut out.

FIGURE 1-10:

Cutting out a rectangle matching the design

7. **Assemble the card.** Now place the plastic assembly inside the card so that the design is centered in the window, with the side of the plastic that you cut facing outward. Tape it in place, including the battery (Figure 1-11).

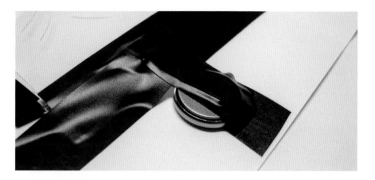

FIGURE 1-11:

Taping the battery in place

With this setup, the card will be lit until the battery goes dead in a week or so, which is fine if you're giving this as a gift in person, but not so good if you're mailing it. For that, you can rig it up so that the LED won't turn on until the recipient opens the card.

8. **Insert a pull tab (optional).** Before taping the battery, insert a paper strip between one of the LED legs and the battery, leaving the end of the strip poking out. This strip will break the circuit and prevent the LED from turning on. Write *Pull* on the end of the strip and tape the battery in place. When the card recipient pulls the paper tab, the leg will again touch the battery, completing the circuit and turning the LED on.

Take It Further

One final trick for extra coolness: use two layers of color with two independent, stacked edge-lit displays, one with a green LED and one with a red LED. You can see an example in Figure 1-12, where a plastic layer with red-lit LEDs is on top of a green layer.

It's not easy to capture with a photo, but the red dots are crisp and clear and float in front of the green tree because of the second layer. The 3D effect is really quite wonderful. We made the dots on the red layer by pressing a push pin into the plastic, rather than using a knife. Inside the card, it's just as you would expect: two layers of plastic with their own LED each at the bottom edge, with electrical tape around the edges to prevent light leakage. This is actually a very old display technique. Stack enough layers, and you can even make a side-lit light-guide numeric display clock!

FIGURE 1-12:

Layer two or more
sheets of plastic on top
of each other to add
multiple colors to the
design.

#3: DARK-DETECTING LED

Here's a common question: How do you make an LED turn on when it gets dark? You might call this the "night-light problem," but it comes up in a lot of familiar situations—emergency lights, street lights, computer keyboard backlights, and the list goes on.

There are many solutions to this problem. The traditional one is to use a circuit with a *CdS photoresistor*, sometimes called a *photocell* or *light-dependent resistor (LDR)*. Photoresistors are reliable and cost only about $1 each, but are being used less and less because they contain cadmium, a toxic heavy metal whose use is increasingly regulated. Luckily, there are many other options.

This project will show you how to design your own dark-detecting LED using a phototransistor, seen in Figure 1-13.

FIGURE 1-13:

This LED turns on when
the lights go out.

You can do almost anything with this inexpensive light-controlled LED circuit, but one fun application is to make LED throwies that turn themselves off in the daytime to save power. Throwies normally can last up to two weeks, and adding a light-level switch like this can significantly extend their lifetime.

Get the Parts

To make this project, you'll need the soldering tools described in the appendix. In addition, you'll use the following electronic components, seen in Figure 1-14.

- Super bright LED (for example, *https://emsl.us/620* or *https://emsl.us/606*; or SparkFun P/N 528 or 10632)

- Coin cell CR2032 battery (SparkFun P/N 338)

- LTR-4206E phototransistor (*https://emsl.us/437* or Mouser P/N 859-LTR-4206E)

- 2N3904 transistor (*https://emsl.us/98* or SparkFun P/N 521)

- 1 kΩ resistor (*https://emsl.us/234* or an assortment such as SparkFun P/N 10969)

The last three parts cost about $0.30 all together, and much less in bulk.

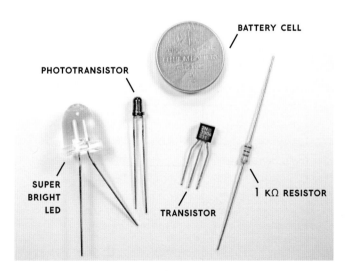

PHOTOTRANSISTOR

BATTERY CELL

SUPER BRIGHT LED

TRANSISTOR

1 KΩ RESISTOR

FIGURE 1-14:
Components needed for your dark-detecting LED

The LTR-4206E is a phototransistor in a 3 mm black package. The black package blocks visible light, so the phototransistor is sensitive only to infrared light—it sees sunlight and incandescent lights, but not fluorescent light or most discharge lamps, which means it can be programmed to only come on at night.

As in the previous projects, you'll start by making an LED throwie. Then, you'll add on the phototransistor and use its output to control the transistor, which turns the LED on.

How It Works

When light falls on the phototransistor, it begins to conduct up to about 1.5 mA, which pulls down the voltage at the lower side of the resistor by 1.5 V. This turns off the transistor, which turns off the LED. When it's dark, the phototransistor is able to conduct about 15 mA through the LED, so the circuit uses only about one-tenth as much current while the LED is off.

Build It

Now it's time to start building. You can certainly put this project together on a breadboard, but there's something more satisfying about the compact and deployable build that we walk through here.

1. **Solder the resistor and transistor.** First get the transistor and the resistor. The pins of the 2N3904 transistor are—from left to right when viewed from the front—emitter, base, and collector. You'll solder the resistor between the leads of the transistor's base and collector, per the diagram shown in Figure 1-15. (If you haven't soldered before or would like some pointers, turn to the appendix.) Hold the resistor with its leads at 90 degrees to those of the transistor and solder, then clip off the excess resistor lead of the transistor base (middle pin) and the collector (right pin).

FIGURE 1-15:

Circuit schematic for the dark-detecting LED

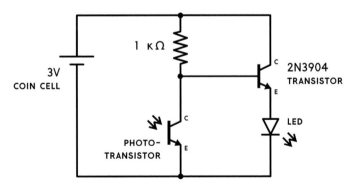

2. **Attach the phototransistor.** Next, add the phototransistor, shown in Figure 1-16. Note that it has a flattened side, much like an LED does. This pin on the flat side is the collector of the

phototransistor. Solder the flat-side collector to the base (middle pin) of the transistor, again at 90 degrees. Leave the other pin of the phototransistor, the emitter, unconnected for now.

3. **Test the LED.** For this step you need to know which LED leg is positive and which is negative. Regrettably, markings on LEDs are not consistent, so the best way to check polarity is to test it with the lithium coin cell: place the LED across the terminals of the cell with one leg on either side. If it doesn't light the first time, swap the side the legs are on. When the LED lights up, note which side is touching the (+) terminal; this is the positive leg (usually it's the one with the longer lead). Solder the positive lead of the LED to the emitter (left pin) of the transistor, which doesn't have anything soldered to it. Trim away the excess lead of the LED that goes past the solder joint. Solder the negative lead of the LED to the phototransistor's emitter—the pin on the non-flattened side, which doesn't yet have anything connected to it. Do not trim this leg.

By this point, there are only two pins extending below the components: one that goes to the resistor and collector (right-most pin) of the transistor, and one that goes to the emitter of the phototransistor and the negative leg of the LED. To test the circuit, squeeze the coin cell between these two terminals, with the positive side touching the resistor. Figure 1-17 shows the circuit in a darkened room; its LED lit up.

FIGURE 1-17:

Darkness triggers the
assembled LED circuit.

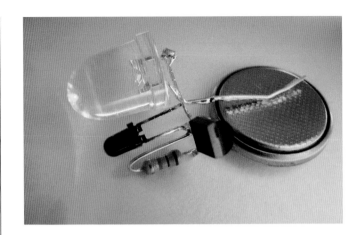

Now you have a tiny dark-detecting light! You can make it more
stable by taping the legs to the battery. This one makes a pretty good
night-light attached to the top of a doorframe—when the room lights
are off, it shines a bright spot on the ceiling.

Take It Further

While this little circuit works on its own, it can also be used as part of
a larger circuit. Certainly, this is one of the easiest and least expensive
ways to control an LED with a photosensor, so you can make several
and adapt them to many purposes, such as combining them with the
sea urchin shells in the first project for urchin night-lights.

#4: ELECTRIC ORIGAMI

The little LED-lit cube seen in Figure 1-18 is much more than just a
paper lantern: it's a translucent thin-film electronic circuit that hooks
a battery up to an LED, and is flexible enough to be folded into an
origami box. The coolest thing about circuits like these? You can
make them at home.

NOTE

*A template for this project
is available in the book's
resources* (https://nostarch
.com/LEDHandbook/).

In this project, you'll combine a basic LED throwie and papercraft, in the form of a traditional origami balloon, to make what
might be called an *LED foldie*. The circuitry consists of aluminum
foil traces, ironed onto adhesive paper (such as freezer paper or
photo-mounting paper) or a laser-printed pattern. This construction
can then be folded to fit an LED and battery to complete the circuit.

FIGURE 1-18:
This LED lantern uses
aluminum foil traces
attached to paper.

Get the Parts

Gather the following parts and materials:

- LED (for example, *https://emsl.us/369* or SparkFun P/N 12062)
- Coin cell CR2032 battery (SparkFun P/N 338)
- Origami paper (we used standard 10 inch squares)
- Aluminum foil
- Freezer paper or photo-mounting paper (the same size as the origami paper)

Build It

The first step in designing a three-dimensional circuit like this is to create the enclosure and see where the parts will go. After that you'll unfold the model, draw circuit paths between the points you want to connect, and go from there.

1. **Fold the balloon prototype.** To get started, first fold an origami balloon. (There are a million tutorials on the web if you need guidance.) This balloon won't be the final one you use, but you'll draw your circuit out on it. Don't inflate the balloon—keep it in its final folded shape, because you'll be transferring the design to a sheet of freezer paper.

    Origami balloons have a couple of features that are useful for this project. They have a convenient pocket on the side, where you'll put a lithium coin cell. They also have a single hole in the bottom; this is where the LED will go, so tuck the leads into the folds on the side opposite the battery. Figure 1-19 shows the balloon with the components temporarily placed.

FIGURE 1-19:

Temporarily place the components on your balloon prototype.

2. **Unfold the balloon.** Mark the locations of the battery and LED terminals on the origami balloon while it's still folded, and then unfold it. This is your circuit board. At this point, you should have the component locations marked, but no lines drawn between them. You can see the unfolded balloon in Figure 1-20.

FIGURE 1-20:

Unfold the balloon to see how your circuit will work.

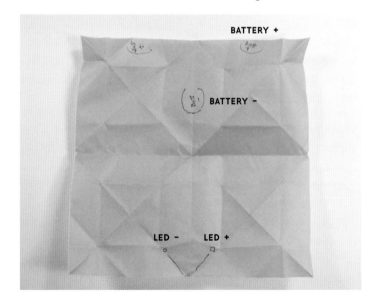

3. **Draw the circuits.** Next, you'll add those circuitry lines (circuit board wires, or traces) between the battery and LED, as seen in Figure 1-21. Draw traces connecting the positive LED leg to the positive battery side, and the negative LED leg to negative on the battery. You'll cut these traces out of foil and stick them inside. One thing to keep in mind for interfacing papercraft to electronics: it's helpful if the circuit traces fold over the leads for the LED in order to maintain good contact, so leaving a little extra at the ends of the traces is always a good idea.

    After connecting the dots (so to speak), you have the resulting layout of your circuit. Pretty simple here—only two traces! The two round pads contact the two sides of the battery, and the two angled pads contact the two leads of the LED.

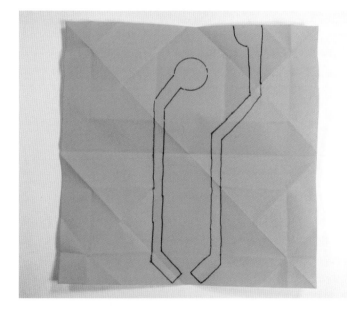

FIGURE 1-21:
Drawing in the traces

4. **Fabricate the circuit board.** The next step is to actually fabricate the circuit board. We call this the *freezer paper* method of attachment, where you laminate foil traces to plastic-coated paper. Lay your circuit diagram over a piece of aluminum foil (Figure 1-22) and trace over it with a stylus (a wooden skewer or blunt toothpick works well) to make an indented outline on the foil.

FIGURE 1-22:

The circuit board
comes together.

5. **Cut out and attach the traces.** Next, cut out your traced pattern. Scissors work well, of course. Be careful not to tear the foil! Figure 1-23 shows the traced pattern, ready to cut.

FIGURE 1-23:

The traced pattern,
ready to cut

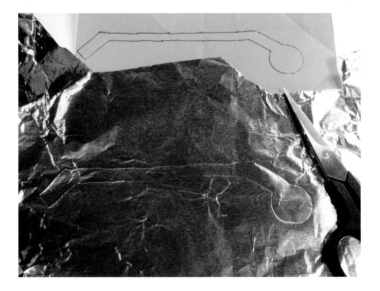

6. **Laminate the foil.** Prefold your freezer paper into the origami balloon shape and compare it to your circuit layout to see where to position the aluminum foil pieces on your freezer paper. Place the foil on the shiny side of the freezer paper, as shown in Figure 1-24. To stop your iron from sticking to the adhesive, place a large sheet of parchment paper over the freezer paper. Once you have your traces positioned, use an iron set to 330 degrees Fahrenheit to laminate the foil to the paper.

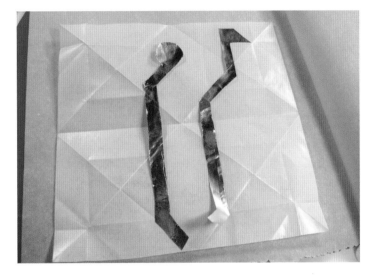

FIGURE 1-24:

Place the foil on the shiny side of the paper.

We used a small hobby iron to fuse the foil to our different papers, but a regular iron works just as well. The dry-mount adhesive does not require much heat, but the iron must be on high to melt the plastic of the freezer paper.

We experimented with waxed paper, which was not sticky enough for the aluminum foil, and even tried applying copper leaf to waxed paper, but it was too fragile and the traces broke upon folding. This combination would probably work reasonably well in an application where folding isn't required: it was absolutely beautiful but completely unreliable for origami. Figure 1-25 shows the traces as the balloon is folded.

FIGURE 1-25:

Folding the balloon
during the ironing
process

HINT

*You won't hurt the LED by
plugging it in backward to
that little battery, so this is
a better method than actually trying to keep track of
the polarity. The LED foldie
naturally wants to sit on the
heaviest part, the battery,
with the LED projecting into
the side of the balloon. The
weight of the battery helps
keep the circuit connected.*

7. **Finish the balloon.** Once the foil is adhered to the paper, refold the balloon with the traces on the outside, inflate it by blowing through the hole at the bottom, and then test the balloon. Insert the components in their correct places, and the balloon should light up. If it doesn't, try turning your battery around. If it still doesn't light up, make sure your LED leads are contacting the traces. It's a friction fit, the strength of which is determined by the tightness of the origami folds.

When you get everything working, it should look like Figure 1-26.

FIGURE 1-26:

The finished balloon

Take It Further: The Direct Toner Method

Our last breakthrough came when we created a PDF pattern to print out. We discovered that you could fuse the foil directly to the toner from a laser printer. This works only for laser printers, not inkjet. You can print out the pattern and iron your foil pieces directly to the paper—the heated toner will adhere to the foil.

There is a caveat with this method. While the foil sticks well to the toner, it isn't quite strong enough that you can just iron on a giant sheet of foil and peel it off so it sticks only where there's toner, so you still need to cut out the foil shapes, at least roughly. After printing out the pattern, place it over your aluminum foil, trace the outlines, and cut out the foil pieces.

Place your foil carefully over the pattern, and iron well at very high heat. Be sure to cover your work with parchment paper, or you'll get toner on your iron. When your foil is stuck to the toner and it's cooled down, cut out the square for the balloon and get ready to fold.

Fold gently so you disturb the traces as little as possible. They may come loose in areas with multiple folds, but should stay on enough for assembly. Inflate the balloon, add the battery and LED, and admire the glow. As before, if you have trouble, try turning your battery around and making sure that the leads of the LED are making contact with the foil.

SUMMARY

In this chapter, you learned about the simplest LED project imaginable, the throwie, and used it in a number of (nonthrown!) projects. You also learned some more challenging techniques that will serve you well, like making your own circuit traces. Hopefully these easy projects have piqued your curiosity about the projects that lie ahead!

THE DESKTOP UFO

BY JAMES FLOYD KELLY

IN THIS PROJECT, YOU'LL BUILD A DECORATIVE UFO THAT LIGHTS UP.

My desk is always covered with half a dozen geeky trinkets—an action figure here, a few old *Star Wars* collectible cards there, and sometimes a new gadget with no real purpose other than to make sounds. Some I keep around for inspiration while writing, and others just add a bit of color and quirkiness to my daily routine. If your desk is lacking in unnecessary trinkets, liven it up with your own version of the Desktop UFO.

The Desktop UFO is lit by LEDs that provide a nice eerie glow. It has an on/off switch to save power and takes up very little space. It consists of a mix of 3D-printed components and items you can likely find around the house, as well as a few electronics components. Including 3D printing time, the entire project can be completed by a beginner in six hours or less.

If you don't have access to a 3D printer, don't worry. A careful examination of the Desktop UFO reveals some very basic shapes. A plastic saucer of any kind could substitute for the 3D-printed version, and the landing pads can be made from plastic or wood or any small, flat objects that can be drilled. It also might be worth checking local libraries and maker spaces to see if there's a 3D printer you can use.

GET THE PARTS

Here is a list of the tools and components you'll need to make the Desktop UFO. This project does require some soldering; if you need instructions, see the appendix.

Components

- Adafruit NeoPixel Ring, 24 × 5050 RGB LED with integrated drivers (Adafruit P/N 1586)
- Adafruit Trinket (Adafruit P/N 1501) and A-to-MicroB USB cable (Adafruit P/N 898)
- Battery holder, 3 × AAA with on/off switch and two-pin JST (Adafruit P/N 727)
- 40 header pins (Adafruit P/N 3002)
- 5- or 7-inch female-to-female jumper wires
- 22-gauge solid wire (Adafruit P/N 1311)
- Plastic lid with at least 1.5 cm of vertical space beneath
- Three bendable plastic drinking straws
- (Optional) Small plastic gumball machine toy enclosure lid
- Spray paint

Tools

- 3D printer or access to a 3D printer

- PLA or ABS filament (depending on 3D printer)

- Drill press, hand drill, and/or rotary tool

- Soldering iron

- Solder

- Wire snips and pliers

- Hot glue gun

- Electrical tape

- Painter's tape

INTRODUCING THE NEOPIXEL

In developing the ideas for the UFO, I found that the Adafruit NeoPixel component, shown in Figure 2-1, offered a nice special effect for the UFO. The NeoPixel is a strip or ring of LEDs that you can program to light up in a pattern, providing a rotating, glowing ring beneath the saucer.

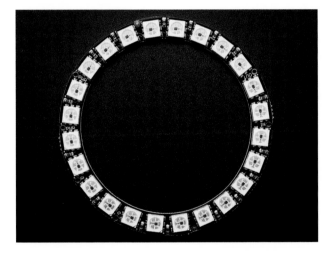

FIGURE 2-1:

The Adafruit NeoPixel

The NeoPixel will be placed beneath the Desktop UFO, with wires pulled through one or two of the legs, which we'll make from drinking straws. The battery case and Trinket controller that controls the lights on the NeoPixel will be situated beneath a round plastic lid recovered from a plastic container. The body of the UFO and the landing pads will be created in a CAD application (Tinkercad) and printed on a 3D printer. Let's get started!

BUILD IT

First we'll set up the Trinket, shown in Figure 2-2. This is the controller board for the whole project that will tell the LEDs on the NeoPixel how they should behave. Then we'll prepare the NeoPixel with a battery and put everything together in the UFO body, which we'll build last.

FIGURE 2-2:

The Trinket board

When everything is built, we'll upload a short program, called a *sketch*, that tells the Trinket the pattern in which to trigger the various LEDs contained in the NeoPixel ring. An LED will light up when voltage is supplied. The NeoPixel provides an "address" for each LED so the sketch can indicate which LEDs the Trinket should provide voltage to at any one time.

Preparing the Trinket

Before you can run the sketch that tells the NeoPixel's LEDs to light up in a particular order, you'll need to complete these steps to prepare your Trinket:

1. **Install the software on your Trinket.** Connect your Trinket to a desktop computer or laptop with an A-to-MicroB USB cable. (There are many varieties of USB cables; the Trinket uses the MicroB. If you do a quick internet search for USB cable types, you can find more details on how to identify the correct cable.)

    Mac and Linux computers don't need extra drivers installed, but if you're using Windows, you'll need to download the drivers. Go to *https://nostarch.com/LEDHandbook/* and follow the link to the drivers for Project 2.

2. **Download the *adafruit_drivers.exe* file.** Run the *adafruit _drivers.exe* installation file (the filename will also include the latest version number) by double-clicking it, then follow the instructions and click through the steps to install all drivers.

   You'll also need to install the Arduino IDE. See "Getting Started with the Arduino and the Arduino IDE" on page 15 for instructions. The following two links will be helpful if you get stuck during the IDE or driver installation: *https://learn.adafruit .com/introducing-trinket/starting-the-bootloader* and *https:// learn.adafruit.com/introducing-trinket/setting-up-with-arduino-ide*.

   You'll need to use the Arduino IDE to allow sketches to be uploaded to the Trinket.

3. **Test the Trinket.** Now we'll set up the Arduino IDE to communicate with the Trinket and use a simple sketch that's installed with the Arduino IDE to test out the Trinket. Make sure your Trinket is connected to your computer and open the Arduino IDE. Go to the Tools menu, select **Board**, and from the list select **Adafruit Trinket 8MHz** as the board type. Again in the Tools menu, click **Programmer** and select **USBtinyISP**. Press the red button on the Trinket to prepare it for the sketch upload. Go to **File ▸ Examples ▸ 01.Basics ▸ Blink**. A new IDE screen will open up with the Blink sketch, as shown in Figure 2-3.

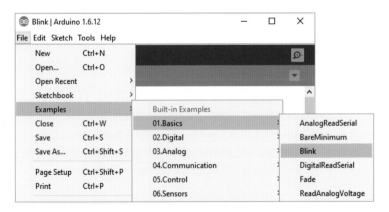

FIGURE 2-3:
The Blink sketch in the Arduino IDE

When the red LED on the Trinket blinks, click the upload button on the Arduino IDE toolbar ( ) to upload the Blink sketch. We'll upload the sketch for this project shortly, but for now we're simply checking that our setup is working.

The red LED on your Trinket should blink over and over. That means it's working! If you don't see the light blinking, try uploading the sketch again, making sure your Trinket is fully connected to the computer.

4. **Upload the NeoPixel sketch.** Our code for controlling the NeoPixel is based on a sketch from Adafruit (*https://learn.adafruit .com/kaleidoscope-eyes-neopixel-led-goggles-trinket-gemma/ software*), with one small change. On this line from the original:

```
for(i=0; i<16; i++) {
```

we changed the 16 to 24 so the line reads:

```
for(i=0; i<24; i++) {
```

The customized sketch, *UFO.ino*, is available in the book's resources at *https://nostarch.com/LEDHandbook/*. Download the file and save it. You'll also need to save the NeoPixel library in your libraries folder. This is also available in the book's resources.

Upload *UFO.ino* into your Arduino IDE by going to **File ▸ Open** and finding the file. This code will cause the NeoPixel to display a number of random animations. If you're good with programming, you can modify the code to do any type of animation you like.

NOTE

Programmers who want to edit their code can use hex code 0xff0000 for red, 0x0000ff for blue, and 0x00ff00 for green.

Once you've uploaded the Adafruit NeoPixel sketch to your Trinket, it's time to add the wiring.

Wiring the Electronics

Next, it's time to wire up and test the electronics before we move to the design of the UFO body.

NOTE

See the appendix for soldering instructions.

1. **Solder the header pins.** The Trinket comes unassembled and requires a little bit of soldering. Included with the Trinket are headers that can be broken apart and soldered to the Trinket.

    For this project, we'll use the following locations on the Trinket:

NOTE

As an alternative to using headers, you can solder four strands of 8-inch wire directly to the small copper-plated holes labeled BAT, 5V, #0, and GND drilled into the Trinket.

    - BAT
    - 5V
    - #0
    - GND

    Solder headers to the Trinket at these locations. You may want to solder additional headers as well, as shown in

Figure 2-4. If you want to add more features to the Trinket later, this will save you from having to solder again.

 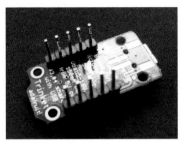

FIGURE 2-4:
Solder headers to the Trinket.

2. **Solder wires to the NeoPixel.** Next, solder connections to the NeoPixel for the following locations:

    - PWR +5V (there are two; either one will work)
    - GND (there are two; either one will work)
    - Data Input

    You can use three headers for this, but I found that using three 8-inch lengths of wire works best, as shown in Figure 2-5. Note that the wires have been inserted up through the holes from the LED side and soldered directly to the small copper rings on the back (black) side of the NeoPixel. It's common convention to use red wires for the power connection and black for GND (ground).

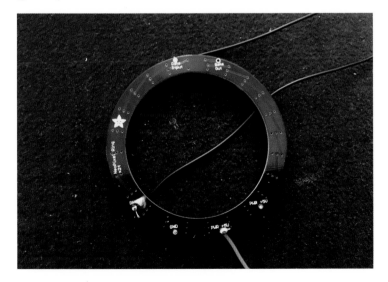

FIGURE 2-5:
Solder wires onto the NeoPixel ring.

NOTE

This step will be much easier if you cut off two female jumper wire ends and solder them onto the red and black wires of the battery case to extend those wires.

FIGURE 2-6:

Create a small shared connection with headers.

3. **Solder the plug.** Next, take your battery case, cut off the white plug on the end, and strip approximately 1/4 inch from the ends of the red and black wires.

    The red wire will provide power to the Trinket, and we'll connect it to the BAT header in a moment. The black wire will not connect directly to the Trinket; instead, the GND header on the NeoPixel and the black wire from the battery case must share the GND header on the Trinket. To make this possible, take the three headers you set aside earlier and apply some solder to all three pins on one side to create a common connection that you can attach three wires to, as shown in Figure 2-6.

4. **Connect up the parts.** Once all the headers have been prepared, make the following connections using the wires you soldered in step 2, or female jumper wires where indicated (see Figure 2-7):

    - NeoPixel Data Input to #0 on the Trinket
    - NeoPixel PWR +5V to the 5V header on the Trinket
    - NeoPixel GND to one of the three common connection headers
    - Battery case red wire to the BAT header on the Trinket
    - Battery case black wire to one of the three common connection headers
    - GND on the Trinket to one of the three common connection headers via female jumper wire

    You can add some heat shrink to cover the solder joints if you have it; if not, simple electrical tape will work fine.

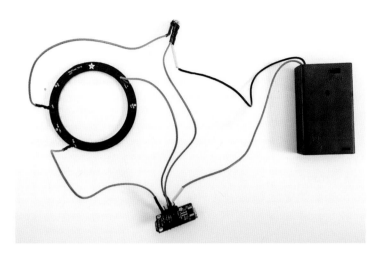

FIGURE 2-7:
All the components wired
up for testing

Once all the connections are made, place batteries in the battery box and move the switch to the on position. If the NeoPixel
sketch was properly loaded on the Trinket and the wiring is correct,
you'll see a random mix of animations and colors displayed on the
NeoPixel.

Troubleshooting

If you're not seeing any animation, double-check all the wiring
connections first. The small three-header common connection is a
frequent culprit, so make sure all wires have a good connection.

If the wiring is good, connect the Trinket to your computer and
upload the sketch again; you do not have to disconnect all the wires
to the Trinket when uploading future sketches. You could also upload
the Blink sketch again and use it to test that the Trinket is functioning
properly.

Once you've got a working NeoPixel displaying the colorful animation, it's time to move on to the UFO body components.

Creating the UFO Body and Landing Pads

Creating a custom UFO body can take some time, so I've included
the 3D printer files I used for the UFO body and landing pads with the
book's resources at *https://nostarch.com/LEDHandbook/*. You can
easily modify these or create your own if you're comfortable with
CAD software.

The UFO body is printed as a single piece. It will be hollow underneath, and this is where the NeoPixel ring will be placed. There will also

be three holes in the body where three straws will be inserted to act as legs.

The body of the UFO was created in Tinkercad (*http://www.tinkercad.com*), as shown in Figure 2-8. It was based on measurements taken from the NeoPixel Ring as well as the diameter of the drinking straws and the base. I also measured the plastic dome so I could add a suitable space for it on top of the UFO body. The measurements are:

NeoPixel   6.5 cm outer diameter, 5.3 cm inner diameter
Drinking straw   0.5 cm outer diameter
Plastic dome   2.9 cm outer diameter
Base   12 cm diameter

FIGURE 2-8:

The UFO body design
in Tinkercad

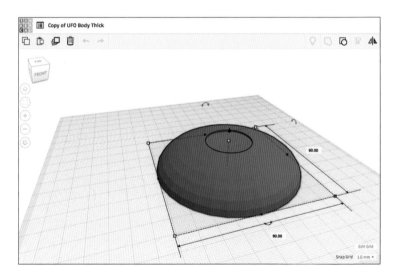

The base for the project is nothing more than a round lid from a large plastic container. You'll need to make sure the lid is at least 1.5 cm tall and large enough to store the battery case beneath without making the entire structure wobble. Remember that the wires will be inserted up through the base and through the straws so they are hidden from view.

We'll also need three small landing pads with holes through the center to run the wires from the Trinket and battery box up through the straws and into the UFO body. Figure 2-9 shows the final design of the landing pads, which can also be created with a 3D printer.

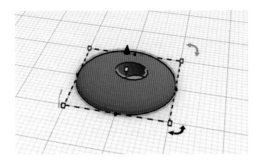

FIGURE 2-9:
The design for the landing pads

Figure 2-10 shows the final printed UFO body, ready for assembly (and later painting).

FIGURE 2-10:
The 3D-printed UFO body

With the electronics ready and the UFO body, landing pads, and base collected, there are just a few final steps to perform before your UFO is ready to go.

ASSEMBLE IT

We'll assemble the Desktop UFO in stages. I recommend reading through this entire section before beginning the assembly so you'll understand the process and how the parts fit together.

1. **Attach the battery box.** Drill a hole in the base of the UFO (the plastic lid) with a 3/8-inch bit, so that the on/off switch on the battery box is accessible. You can use a location of your own choosing, but I elected to drill the hole near the edge of the base, as shown in Figure 2-11.

FIGURE 2-11:

Hole for accessing the
on/off switch

Position the battery box so you'll be able to change out the
batteries. Before gluing, make sure the on/off switch is aligned
with the hole you drilled for it. Apply a small bit of hot glue to the
back of the battery box and attach it to the bottom of the base.
Figure 2-12 shows the battery box glued in place.

FIGURE 2-12:

Hot-glue the battery
box to the bottom of
the base.

Once the glue has dried, flip the base over and ensure
that you can move the on/off switch through the drilled hole as
shown in Figure 2-13.

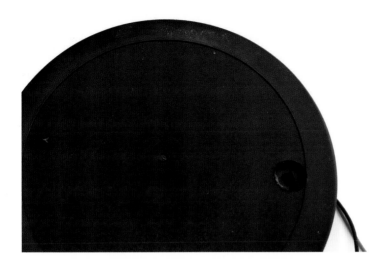

FIGURE 2-13:
The on/off switch should
be accessible.

2. **Attach the NeoPixel.** Next, use two small dabs of hot glue
    to secure the NeoPixel to the underside of the UFO body.
    Figure 2-14 shows that I've used hot glue on opposite sides of
    the NeoPixel; if you let the hot glue cool for just a few seconds
    before placing it into the body, it shouldn't damage the NeoPixel
    and will allow you to center it under the UFO body.

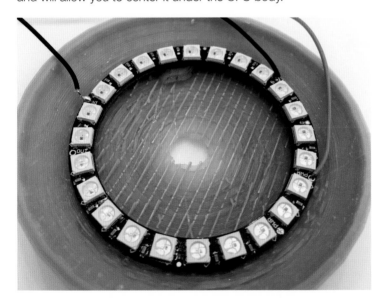

FIGURE 2-14:
Hot-glue the NeoPixel ring
beneath the UFO body.

Use blue painter's tape to keep the three wires together at
the edge of the body. You should also use small bits of tape to
label each of the wires from the NeoPixel (GND, PWR +5V, and
Data Input).

3. **Prepare the landing pads.** Place a bit of hot glue around the bendable portion of each drinking straw and insert the short, flexible end into each landing pad as shown in Figure 2-15. Don't insert the straw into the hole too far; just enough for the glue to adhere the straw to the landing pad is sufficient. There should be a little bit of the flexible portion of the straw visible above each landing pad.

FIGURE 2-15:

Hot-glue the flexible section of each straw to a landing pad.

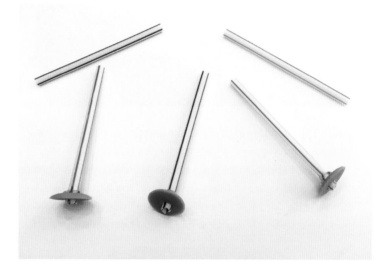

When you've finished gluing the flexible ends into the landing pads, cut off about 3 inches from the longer length of each straw; you can fine-tune the length once you get the UFO assembled, but 3 inches or more should give you enough straw to experiment with as you balance the saucer on the legs.

4. **Secure the Trinket and drill the hole for the landing pad.** The Trinket will need to be secured beneath the base; you can use painter's tape to temporarily attach it until your Desktop UFO is completed and then secure it with hot glue or stronger tape.

Figure 2-16 shows that I've secured the Trinket and organized the wires so they are close to the location where I will drill a hole for one of the landing pads.

You only need to drill a hole under one of the landing pads. That hole can also be drilled with a 3/8-inch drill bit (and you can drill it when you drill the on/off hole described earlier). The flexible straws will allow you to bend the legs, so you don't have to be exact with where you drill this hole, but try to imagine an equilateral triangle where the landing pads will be glued beneath the UFO body and drill a hole for the landing pad that will be farthest from the battery box on/off switch.

FIGURE 2-16:
The Trinket is secured and the wires bundled together.

You will have to disconnect some wires to feed through the straws; Figure 2-17 shows the hole drilled for one of the landing pads and the wires from the Trinket pulled through.

FIGURE 2-17:
Landing pad hole and wires from the Trinket

5. **Connect the wires from the Trinket to the NeoPixel.** Using one of the legs you made with the landing pad and straw, insert all three wires from the Trinket so the ends come through the straw. If the straw is too long, trim it down until all three wire ends can be easily connected to the NeoPixel wires as shown in Figure 2-18.

FIGURE 2-18:

Trinket wires connected to NeoPixel wires

Remember, connect the 5V wire from the Trinket to the PWR +5V wire on the NeoPixel. (In the photo, that wire is red, but your wire color scheme could be different.) The GND wire from the Trinket connects to the GND wire on the NeoPixel, and the #0 wire on the Trinket connects to the Data Input wire on the NeoPixel.

After all the wires are connected, flip the on/off switch and test that the NeoPixel still displays animation. If not, recheck all your connections against the preceding instructions and turn back to "Troubleshooting" on page 55.

6. **Glue the landing pads.** Place a small bit of hot glue on the bottom of each landing pad; you may have to trim off more plastic from the straw to get the pad to sit flat on the base. Figure 2-19 shows all three landing pads glued to the base.

You can adjust the angle of the landing pads using the flexible sections of the straws. Angle the landing pads so that the UFO body rests on top of the three legs. Place the clear dome on top and turn the switch on. Hopefully you'll see a nice glow from beneath the UFO!

FIGURE 2-19:
Hot-glue the three landing
pads to the base.

If you're happy with the animation, you can use some additional
hot glue to secure the UFO body to the legs; be careful when turning
the base and legs over and ask for a second set of hands to help
you get the legs glued in place properly. Tuck away any exposed
wire using tape or a dab of hot glue applied to the underside of the
UFO body.

Finish up with some paint (I recommend silver for the UFO body)
and you'll end up with a cool-looking desk sculpture.

TAKE IT FURTHER

This UFO can easily be modified. You could add some weapons or
maybe change the shape to something more menacing. You can
replace the NeoPixel with some of Adafruit's other specialty LEDs—
there are so many variations, you're likely to find something that
grabs attention.

Here are some other ideas to upgrade the Desktop UFO:

- Add a pulsing LED beneath the clear dome
- Add random sound effects
- Put a toy cow figurine beneath the UFO

SUMMARY

There are a lot of ways to embellish the Desktop UFO, and this really
is a great project to teach a young student some basics in electricity
and programming as well as prototyping. The final UFO doesn't have
to look perfect; this one has its flaws, but it works and looks cool—
and that's good enough for me!

THE CHEATER'S DICE ROLLER

BY JOHN BAICHTAL

IN THIS PROJECT, YOU'LL BUILD A DIGITAL DICE ROLLER WITH A TWIST.

Do you suffer from polyhedral dice elbow from playing too much *Dungeons & Dragons* or other tabletop role-playing games? (I'm talking about the dice with lots of different shapes. A player might strain a ligament throwing those dice so often!) Or maybe you just want the convenience of generating your die rolls electronically? Either way, this dice roller is the solution, and then some.

You have the option to switch between a D&D standard 20-sided die and the two 10-sided dice of *Warhammer Fantasy Roleplay* and other systems, though I'll show you how to customize this to any two-digit variety. However, this is no ordinary dice roller: built-in cheats set it apart. I'll show you how to trigger a high roll or a low roll without any of your hapless gaming friends ever knowing, using nothing but magnets, sensors, and some sleight of hand.

I built the project enclosure as a medieval castle made out of laser-cut wood. The last section of this chapter will talk more about the idea behind my enclosure and ways to make your own.

GET THE PARTS

Grab the following parts to build your dice roller. I ordered most of the parts from Adafruit and SparkFun.

Components

NOTE

See "Getting Started with the Arduino and the Arduino IDE" on page 15 for setup instructions.

- Arduino Uno (Adafruit P/N 50 or SparkFun P/N 11021)
- USB A-B cable (Adafruit P/N 62)
- Two 8 × 8 LED matrices with I^{2}C backpacks (Adafruit P/N 872)
- Two reed switches (SparkFun P/N 8642)
- Large push button (for example, SparkFun P/N 9336)
- Double-throw switch (Adafruit P/N 805)

NOTE

Power the project with either a 9 V wall wart (such as Adafruit P/N 63) or a 9 V jack adapter (such as Adafruit P/N 80) plugged into the power jack on the Arduino.

- Wall wart or 9 V battery clip (Adafruit P/N 63 or Adafruit P/N 80)
- Piezo buzzer (Adafruit P/N 1739)
- 10 kΩ resistors (SparkFun P/N 10969 is a good multipack)
- Magnet, powerful enough to trigger the sensor reliably (for example, Adafruit P/N 9)
- (Optional) Breadboard (SparkFun P/N 12002)

Tools

- Laser cutter or saw (such as a jigsaw or band saw)
- Soldering iron
- Solder

- Wire snips and pliers
- Wire
- Hot glue gun
- Wood glue
- Spray paint

INTRODUCING THE LED MATRIX

We're using 8 × 8 LED matrices to display the results of the dice throws. Every pixel of an LED matrix is independent, and you selectively trigger them, lighting up or dimming each LED to make a pattern. The simplest way to do this is to store the status of each pixel in an array. For instance, Figure 3-1 shows an array of LEDs selectively lit to make a smiley face.

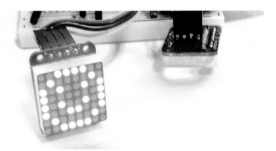

FIGURE 3-1:
Use arrays to store smiles.

You can control your LED matrix with code on an Arduino. The Arduino sketch looks at the array to see the status of each LED, lighting or turning them off every time the sketch loops.

The following code snippet gives you a sense of how it works:

```
smile_bmp[] =
{ B00111100,
  B01000010,
  B10100101,
  B10000001,
  B10100101,
  B10011001,
  B01000010,
  B00111100 },
```

Each line represents a row on the matrix, and each digit represents one LED. When the LED value is set to 1, it's lit. When it's set to 0, it's unlit. You can design any graphic you want with this method.

However, typing this out for every design can be tedious, so there's a convenient database of shapes controlled by an Arduino library for the matrix used in this project. The library stores functions that draw letters and numbers, as well as basic shapes like squares and circles, so you don't need to design these shapes from scratch. In "Code It" on page 71, I'll explain how the sketch interacts with the library to draw those default shapes.

BUILD IT

Follow these steps to build a dice roller that can secretly do your bidding:

NOTE

If you need instructions on soldering, see the appendix.

1. **Solder the matrices to their boards.** The matrices come with mini control boards known as *backpacks* that manage the complexity of running 64 LEDs with just a few wires. You need to solder these backpacks to your matrices: add the matrix to the backpack with the IC (*integrated circuit*, also known as a *microchip*) on the underside and solder the pins in place in the holes. The end result should look like Figure 3-2. There is no up or down to the matrix, as long as you add it to the correct side of the board. If you need more guidance, refer to the page for this product (P/N 872) on *http://www.adafruit.com/*.

FIGURE 3-2:
Solder in one or two of these jumpers to change the matrix's I²C address.

2. **Solder in pins or wires.** There are four pins that connect the matrix to the rest of the project. If you plan to use the matrices with a breadboard, solder in the accompanying header pins. If you want to use wires instead, leave them off.

3. **Solder board pin A1.** Of particular importance are the two A0 and A1 solder pads, seen in Figure 3-2, which allow you to daisy-chain up to four matrices by selectively soldering the

pads. This is known as "changing the I^{2}C address." If you solder none of the pads, the I^{2}C address for that matrix defaults to 0x70. Soldering A0 sets the value to 0x71, soldering A1 sets it to 0x72, and soldering both A0 and A1 identifies the matrix as 0x73. Giving each matrix its own address allows us to talk to one matrix without the other responding even though they share wires. You need to add solder to just one of these pins on one board, so add solder to pin A1.

4. **Connect the matrices to the Arduino.** Add your matrices to a breadboard and connect the power and ground rails to 5V and GND, respectively, and then connect both power rails at either side of the board, as shown in Figure 3-3. Connect power and ground of the first matrix to power and ground on the board, then connect up the ground and power pins of both matrices so they are both powered. You'll also need to connect the data and clock wires. Connect pins A4 and A5 on the Arduino to the D and C pins, respectively, on one backpack, and with another pair of wires connect D and C of the first matrix to D and C of the second matrix. In Figure 3-3 the data and clock wires are yellow and green, respectively.

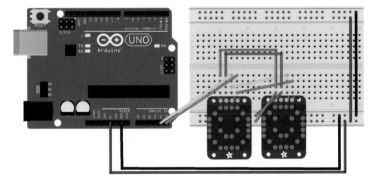

FIGURE 3-3:
Attach the matrices.

Depending on your project, you may need to connect the matrices with wires rather than plugging them into the breadboard in order to make them fit. If this is the case, you can still attach the matrices as shown in Figure 3-3.

5. **Attach the double-throw switch.** Next, connect the double-throw switch, as shown in Figure 3-4, with the middle lead connected to 5V (pink wire) and the left and right leads connected to digital pins 10 and 11 on the Arduino, respectively (brown wires). Also connect the left and right leads of the switch to ground via 10 kΩ resistors, shown in white in Figure 3-4.

This switch will determine whether this is in D&D d20 mode or Warhammer d100 mode: the Arduino can check the position of the switch by scanning pins 10 and 11.

FIGURE 3-4:

Adding the double-throw switch

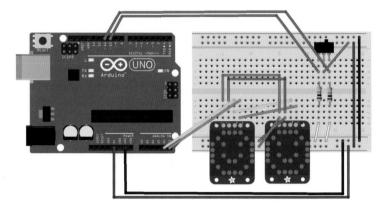

6. **Connect the reed switches.** The reed switches come next. As shown in Figure 3-5, connect the switches to pins 5 and 6 on the Arduino (purple wires) with the other leads going to ground (gray wires).

FIGURE 3-5:

The reed switches tell the Arduino when a magnet is near.

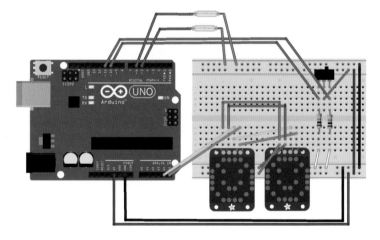

7. **Add the reset button.** The reset button isn't for starting over when you mess up; it's really your roll button. The project software dictates that the dice roll once and then stop, and you must cycle the power or reset the Arduino in order to reroll. Figure 3-6 shows the button installed. Connect one lead to Reset on the Arduino's power bus (orange wire) and the other to GND (pink), soldering in wires to the button's leads as needed. When the button is pressed, the Arduino restarts and automatically reruns the dice roller program.

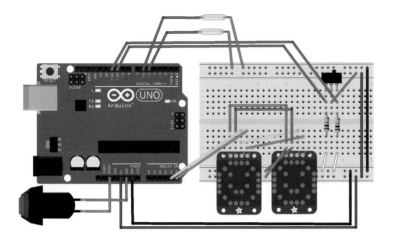

FIGURE 3-6:
Adding a button to reset
the Arduino

8. **Install the buzzer.** The buzzer completes the project. It makes a sound to signal the completed die roll. The buzzer I included in the parts list features breadboard-friendly leads and is attached as shown in Figure 3-7: connect the positive lead to pin 9 on the Arduino and the negative lead to GND.

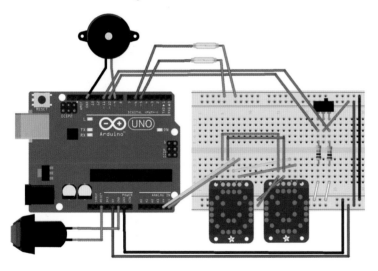

FIGURE 3-7:
Wiring up the buzzer

Now it's time to add code!

CODE IT

Now that the electronics are assembled, it's time to test the project by uploading code to the Arduino.

The LEDBackpack library does most of the heavy lifting in this script. The backpack's library has code for the full alphabet, as well as to draw shapes like lines, circles, and squares. You'll need to download the library from the Adafruit product page (*https://learn .adafruit.com/adafruit-led-backpack/0-8-8x8-matrix*) and save it to your Arduino libraries folder before you can make use of it. The product page also includes tutorials and tons of information if you want to learn more.

The code for the project is also in the book's resources, so open that in the Arduino IDE now and let's go through the code section by section.

The first part of the code consists of references to three libraries used in the sketch.

```
#include <Wire.h>
#include "Adafruit_LEDBackpack.h"
#include "Adafruit_GFX.h"
```

Next, the backpacks are formally initialized:

```
Adafruit_8x8matrix matrix1 = Adafruit_8x8matrix();
Adafruit_8x8matrix matrix2 = Adafruit_8x8matrix();
```

We name the backpacks matrix1 and matrix2. We then enable serial communications for debugging purposes:

```
void setup() {

  Serial.begin(9600);
```

This allows us to communicate with the Serial Monitor. Next, we initialize some pins:

```
  pinMode(5, INPUT_PULLUP);
  pinMode(6, INPUT_PULLUP);
  pinMode(9, OUTPUT);
  pinMode(10, INPUT);
  pinMode(11, INPUT);
```

The pins interacting with the double-throw switch, the buzzer, and the reed switches are initialized. In the latter case the pins are initialized so as to use the Arduino's built-in resistors, ensuring no false positives trigger the cheat mode.

This line seeds the random number:

```
  randomSeed(analogRead(0));
```

By taking a reading from pin A0, we seed the random number we'll use as our dice throw. Then we need to start up the matrices:

```
matrix1.begin(0x70);  // the default
matrix2.begin(0x72);  // with A1 soldered

matrix1.clear();
matrix2.clear();
```

This lets the Arduino know to power up the matrices, and makes sure all of the LEDs begin as off. These are the pin declarations for the switches:

```
int reedLow = 5;
int reedHigh = 6;
int switch1 = 10;
int switch2 = 11;
```

Remember we have two settings for different types of dice. This is the number generation functionality for the d100 dice:

```
// switch selection roll d100

if (digitalRead(switch1) == HIGH) {

  int tensDigit = random(0, 10);
  int onesDigit = random(0, 10);

  Serial.print(tensDigit);
  Serial.println(onesDigit);
  Serial.println();
```

We'll use the throw switch to switch between d100 mode and D&D dice mode. Here, if the switch is set for d100, the Arduino rolls two random numbers, each between 0 and 9. Then it prints them to the Serial Monitor—but doesn't print them to the matrices yet.

We then need to listen to the reed switches to see if they have been tripped by a magnet using two if statements—this is our cheat.

```
if (digitalRead(reedLow) == LOW) {
  tensDigit = 0;
  onesDigit = 1;
}

if (digitalRead(reedHigh) == LOW) {
  tensDigit = 0;
```

```
    onesDigit = 0;
  }
```

If a reed switch is tripped, the random value rolled up is superseded by either a maximum or minimum roll, depending which reed switch has been tripped. The two numbers are written to `matrix1` and `matrix2`:

```
  // write result
  matrix1.setTextSize(1);
  matrix1.setTextWrap(true);
  matrix1.setTextColor(LED_ON);
  matrix1.print(tensDigit);
  matrix1.writeDisplay();

  matrix2.setTextSize(1);
  matrix2.setTextWrap(true);
  matrix2.setTextColor(LED_ON);
  matrix2.print(onesDigit);
  matrix2.writeDisplay();
}
```

Then we have the random number generator for the d20:

```
// switch selection roll d20

if (digitalRead(switch2) == HIGH) {
  int d20result = random(0, 20);

  Serial.print("d20 result: ");
  Serial.println(d20result + 1);
  Serial.println();
```

If the switch is set to d20, the classic D&D die, the Arduino generates a random number between 0 and 19, adds 1, then sends the result to the Serial Monitor. The following series of `if` statements allows the sketch to supersede the rolled number with the cheats if a reed switch is tripped:

```
  if (reedLow == LOW) {
    d20result = 0;
  }

  if (reedHigh == LOW) {
    d20result = 19;
  }
```

```
if (d20result == 19) {
  tensDigit = 2;
  onesDigit = 0;
}

else if (d20result > 9) {
  tensDigit = 1;
  onesDigit = d20result - 9;
}

else if (d20result == 9) {
  tensDigit = 1;
  onesDigit = 0;
}

else if (d20result < 9) {
  tensDigit = 0;
  onesDigit = d20result + 1;
}

}
```

Whatever the outcome, the number rolled needs to be displayed on the matrices:

```
// write result
matrix1.setTextSize(1);
matrix1.setTextWrap(true);
matrix1.setTextColor(LED_ON);
matrix1.print(tensDigit);
matrix1.writeDisplay();

matrix2.setTextSize(1);
matrix2.setTextWrap(true);
matrix2.setTextColor(LED_ON);
matrix2.print(onesDigit);
matrix2.writeDisplay();
```

This displays the result. The program runs once and then stops.

```
}

void loop() {
}
```

Because we only want the code to run once, there is no need for a loop and it goes unused! The sketch is complete.

ASSEMBLE IT

Next, you'll need a box of some sort that will enclose your project. You can approach this step in a few different ways.

Laser-Cut Your Own Enclosure

My first suggestion for anything usually involves designing and creating it yourself. My design resembles a small castle (see Figure 3-8), and this not only gets you in the mood for swordplay and magic spells, but also helps disguise the way you trigger the reed switches to swing the game your way. (More on that last bit after this section.) Here I'll show you how I went about building my castle enclosure.

FIGURE 3-8:

My castle-slash-dice roller, ready for action

1. **Design the case.** Design your case, using either a vector art program like Inkscape (*https://inkscape.org/*) or an online tool like MakerCase (*http://www.makercase.com/*) that designs the case for you. You can also download my castle design in the book's resources (*https://nostarch.com/LEDHandbook/*) and re-create or modify it as you see fit. Figure 3-9 shows my case in Inkscape. Cut it out of quarter-inch plywood or the equivalent— I used a double-layer of eighth-inch plywood.

2. **Output the design.** Cut out the design on a CNC router or laser cutter. Figure 3-10 shows my design fresh from the laser cutter.

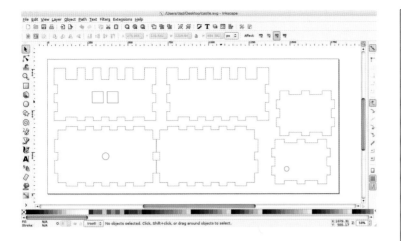

FIGURE 3-9:
I designed my castle design in Inkscape with help from *makercase.com*.

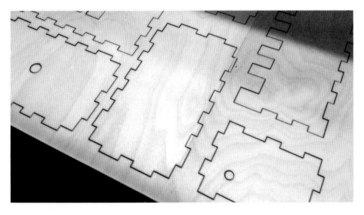

FIGURE 3-10:
My design, freshly lasered

3. **Assemble the castle.** Clean up and assemble the castle, as shown in Figure 3-11. Few designs come out of the machine ready to assemble, so you should plan to work on the cut parts with files and sandpaper to get them to fit together. When you're ready, glue the parts together.

FIGURE 3-11:
Assemble the lasered parts.

4. **Paint and decorate the castle.** Paint the castle, as shown in Figure 3-12. I suggest glossy spray paint to ward off dirt and fingerprints. Another good source of paint (given the subject matter) is your local hobby store, which doubtlessly offers innumerable shades of "dungeon gray" and "slime green" for adding details. What's a proper dice roller without bloodstains and moss?

5. **Add electronics.** Place your project inside the enclosure, making sure the button is accessible. The arcade button has a deep footprint and may interfere with the Arduino's placement. You also need to be mindful of the placement of the reed switches. Be sure to test the switches' placement with a magnet before securing them to determine where you need to place the magnet so they can sense it.

Alternate Enclosures

If you don't have access to a laser or mill, you'll have to find something else to contain your project. Here are a couple of suggestions:

- **Repurpose another box.** Simply find a sturdy box the right size to hold the components, and put them inside. Cut holes so the button, switch, and matrices can be installed.

- **Buy an enclosure.** You can also find a plethora of inexpensive project boxes on the web. One of my favorite sources is Jameco Electronics (*https://www.jameco.com/*), which offers different sizes and levels of durability. As with the repurposed box, you'll need to be aware of how much space the project's guts will take up. An 8 × 6 × 3 project box like Jameco P/N 18869 does the trick.

- **Build one out of LEGO bricks.** Chances are, you already have a great enclosure in your house, kept in a bucket in the basement next to the holiday boxes. I'm talking about LEGO! Simply build a box the right size, leaving holes for the switch, button, LED matrices, and power cord.

USE IT

Simply power the dice roller by plugging your wall wart or 9 V power adapter into the power jack, and press the button every time you want to roll a die, using the magnet on either reed switch depending on your diabolical goals.

I disguised my magnet by taking a gaming figure with a hollow plastic base and hot-gluing the magnet into the base, as shown in Figure 3-13. The reed switches are positioned close to the top of the castle, and to trigger them you simply place the magnetized figure next to the battlements along with some innocuous figures. During the course of the game you move the figures around the "castle," making sure to move the triggering figure to the correct spot as needed.

FIGURE 3-13:
Hot-glue a magnet to the base of a gaming figure and use it to trigger your dice roller.

SUMMARY

When you complete the hardware and software portion of this project, it should look something like the photo on page 65, or a variation thereof with your own imaginative design.

Hopefully, you're also wrestling with the morals of hoodwinking your gaming associates. Good luck with this project!

COLOR-CODED LED WATCH

BY JOHN BAICHTAL

IN THIS PROJECT, YOU'LL MAKE A WATCH THAT DISPLAYS THE TIME USING COLOR.

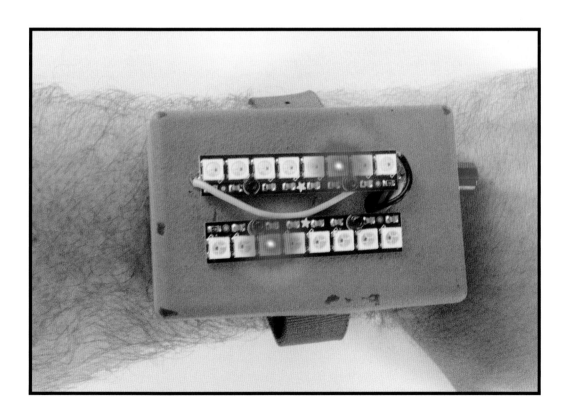

With digital clocks found on phones, microwaves, and even coffee makers, you might think making another is redundant. On the other hand, how often have you needed to know the time while on an important phone call, on a hike, or just nowhere near your kitchen? There is still a place for the portable watch, and to modernize it I've come up with a solution: a custom digital timepiece featuring an intriguingly cryptic display.

With this project you'll make your own watch using a pair of eight-dot NeoPixel LED strips to convey the time with *colors*. It may not be immediately clear how you'd tell the time with colors, but stick around and you'll find out. The brains for our watch will be an Arduino, which simplifies certain parts of displaying the time but also presents special challenges.

GET THE PARTS

You'll need the following parts to build this project.

Components

NOTE

See "Getting Started with the Arduino and the Arduino IDE" on page 15 for setup instructions.

- Arduino Pro Mini (SparkFun P/N 11113; extra headers are available as P/N 00116.)

- FTDI Basic Breakout (SparkFun P/N 09716; this is a programming dongle necessary to program the Arduino Pro Mini.)

- 2 NeoPixel strips (Adafruit P/N 1426 or SparkFun P/N 12661; this is an eight-LED RGB strip.)

- DS1307 RTC Breakout Board (SparkFun P/N 12708; other options are Adafruit P/N 255 or Adafruit P/N 3013.)

- 470 Ω resistor (SparkFun P/N 10969 is a good multipack)

- Enclosure (My enclosure was 85 mm × 55 mm on its largest side and 30 mm deep; find it by searching online for its P/N, WCAH2855. A similar part is the classic 3 × 2 × 1 project enclosure from RadioShack, P/N 2701801.)

- Watch band (Any generic nylon or rubber watch band should do the trick. I used a knockoff iWatch band from Amazon.)

- 9 V battery pack and battery (A standard 9 V connector with wire leads; I'm using Jameco P/N 109154. You can also get the SparkFun P/N 00091 snap connector and cut off the plug.)

- Button (I used a panel-mount momentary button from SparkFun, P/N 11992. The smaller, the better. You can swap out the button for a switch if you want to keep the display going until you turn it off.)

- Screws (I used M2 × 10 mm screws and nuts from HobbyKing, P/N HA0506.)

The small Arduino we use here offers the same Arduino experience without all the bells and whistles so that it will fit in a small enclosure. For instance, you can't program it via USB, and you'll need a FTDI Breakout (such as SparkFun P/N 09716) to program it. Before you begin building this project, see "Arduino Boards Without a Built-in USB Connection" on page 21 for information about how to set up the Arduino Pro Mini and FTDI.

When choosing an enclosure, keep in mind that the LED strips and the 9 V battery are the main limitations on dimensions. You may also want a clip to keep the battery from rolling around inside the enclosure box, though mine was so densely packed nothing could move around.

Tools

- Soldering iron and solder
- Dremel or similar rotary tool, with cutting and drilling implements
- (Optional) Spray paint

INTRODUCING THE REAL-TIME CLOCK MODULE

It turns out that keeping accurate time isn't something Arduinos do well. They can keep track of time from second to second fairly well, thanks to the timing crystal built into the board, but they must use their own internal memory to retain this information, so when the memory fills up your clock stops working. Another problem arises when the Arduino loses power—you lose the time and have to reset it when you start the Arduino up again. The solution to these challenges lies in an add-on board called a *Real-Time Clock (RTC)* module.

The RTC consists of a dedicated timekeeping chip and a battery backup, so it can retain the time even when the main project powers down. It keeps track of the time so your Arduino doesn't have to.

Figure 4-1 shows the DS1307 RTC module sold by Adafruit Industries. It accurately computes calendars up to the year 2100 with leap year factored in, and communicates with the Arduino through a two-wire interface.

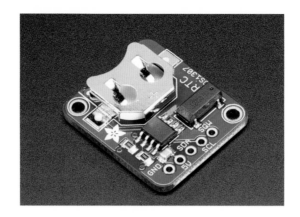

FIGURE 4-1:

Adafruit's DS1307 breakout board helps your Arduino keep time.

HOW IT WORKS: COLOR-CODING THE LEDS

With only a small watch face to utilize, we are limited to just two eight-LED strips to show the time. I devised a system—shown in Figure 4-2—that uses the color of the LEDs to convey the information. I selected five colors to stand in for numbers, listed in Table 4-1.

TABLE 4-1:

Number codes for the colors

COLOR	VALUE
Black	0
Blue	1
Red	2
Yellow	4
Green	5

The top LED strip will indicate the hour, and the bottom strip the minutes. As mentioned, each strip will have eight LEDs. Each LED's position in that strip will be the *multiplier* that you apply to the number associated with the color in Table 4-1. For example, the seventh LED lit up yellow (4) would be 4 × 7, so would mean 28 minutes if it's lit in the bottom strip.

You could use more or fewer colors, but I decided to use only four (plus black) for simplicity's sake. I wanted my clock to use the same colors manufacturers use to color-code resistors—in their world, 0 is black, 1 is brown, 2 is red, and so on. However, brown doesn't show up well as visible light, so I substituted blue. Ultimately it doesn't matter what colors you choose for each time. If it makes sense to you, go for it.

Listing 4-4 consists of the `loop()` function, which contains the code that controls the activation of the LED strips.

LISTING 4-4:

The activation loop

```
void loop() {
    //determine time
    DateTime now = RTC.now();

    int hourDisplay = (now.hour(), DEC);
    int minuteDisplay = (now.minute(), DEC);

  Serial.println();
    Serial.print("time: ");
    Serial.print(hourDisplay);
    Serial.print(':');
    Serial.println(minuteDisplay);

  //hour display
  for (int thisPin = 0; thisPin < 8; thisPin++) {

    int LEDcolor = numberMatrix[hourDisplay][thisPin];

    switch (LEDcolor) {

      case 0:
        strip.setPixelColor (thisPin, 0, 0, 0);
        break;

      case 1: //make that pixel blue
        strip.setPixelColor (thisPin, 0, 0, 100);
        break;

      case 2: //make that pixel red
        strip.setPixelColor (thisPin, 100, 0, 0);
        break;

      case 4: //make that pixel yellow
        strip.setPixelColor (thisPin, 100, 50, 0);
        break;

      case 5: //make that pixel green
        strip.setPixelColor (thisPin, 0, 100, 00);
        break;

    }
  }
```

```
    for (int thisPin = 0; thisPin < 8; thisPin++) {
      int LEDcolor = numberMatrix[minuteDisplay][thisPin];

    switch (LEDcolor) {

      case 0:
        strip.setPixelColor ((thisPin+8), 0, 0, 0);
        break;

      case 1:
        strip.setPixelColor ((thisPin+8), 0, 0, 100);
        break;

      case 2:
        strip.setPixelColor ((thisPin+8), 100, 0, 0);
        break;

      case 4:
        strip.setPixelColor ((thisPin+8), 100, 50, 0);
        break;

      case 5:
        strip.setPixelColor ((thisPin+8), 0, 100, 00);
        break;

    }

  strip.show();

  }

  Serial.println();
  delay(1000);
}
```

The sketch launches the instant you power up the Arduino
(by pressing the button or hitting the switch). The loop reads the
time from the RTC and then lights up the LEDs with whatever color
is appropriate, displaying them as long as the board is powered.
Each LED has a number (0–7 for hours and 8–15 for minutes), and
receives color data from the array at the beginning of the sketch. If
you use a switch instead of a button, it will keep the LEDs lit continuously, and the time will update every second.

SUMMARY

Once you complete the physical build and upload the code, your watch is done! It may not be a precision instrument, but it will certainly start a conversation or two.

5

REAL-TIME MUSIC VISUALIZER

BY MICHAEL KRUMPUS

THIS PROJECT SHOWS YOU HOW TO CREATE AN AWESOME MUSIC VISUALIZER USING AN ARDUINO AND AN LED STRIP.

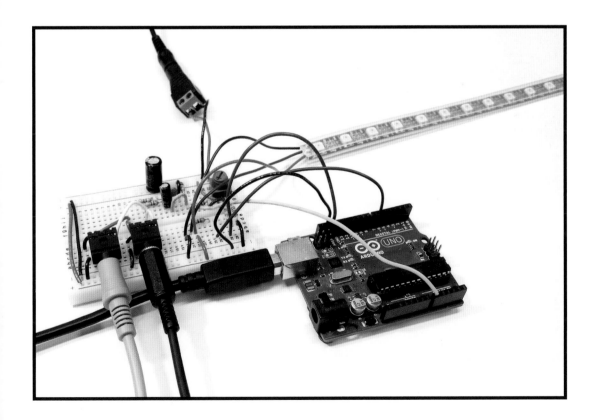

After building this project, you'll be able to plug any music source into the circuit and the LED strip will display a cool, multicolored light show synchronized to your music in real time. This project is great for parties, DJ booths, or even your dorm room! To see the visualizer in action, have a look at this short video: *https://youtu.be/GkjBT-EmRw8*.

First we'll look at the theory behind making a music visualizer so you have some understanding of how it works before getting into the build. After the circuit is built, you'll learn how the code works and how to load it into the Arduino. Finally, we'll cover some tips on how to revise the code to make it behave differently so you can experiment on your own.

GET THE PARTS

The parts for the Real-Time Music Visualizer are easy to find. This list suggests Adafruit for some of the more specialized parts and Mouser for simple passive components.

NOTE

The length of your LED strip determines your power needs. For a 2 m strip, a 2 A power supply is enough. But if you use a 3 m strip, you'll need a 5 A or 10 A supply. Adafruit P/N 658 is a good option for 10 A.

- Arduino Uno (Adafruit P/N 50; see "Getting Started with the Arduino and the Arduino IDE" on page 15 for setup instructions)
- 1, 2, or 3 m WS2812B RGB LED strip (for example, Adafruit NeoPixel P/N 1461)
- 5 V, 2 A power supply (Adafruit P/N 276, or P/N 658 for 10 A)
- 2.1 mm DC breadboard power jack (Adafruit P/N 368)
- Two breadboard audio jacks (Adafruit P/N 1699)
- 10 kΩ potentiometer (for example, Adafruit P/N 356)
- Solderless full-size breadboard (Adafruit P/N 239)
- 22-gauge solid hookup wire for connecting components (Adafruit P/N 1311)
- Three 4.7 kΩ resistors (Mouser P/N 291-4.7K-RC)
- Two 100 kΩ resistors (Mouser P/N 291-100K-RC)
- 2.2 kΩ resistor (Mouser P/N 291-2.2K-RC)
- 470 Ω resistor (Mouser P/N 291-470-RC)
- 0.047 μF (47 nF) ceramic capacitor (Mouser P/N 594-K473K10X7RF5UH5)
- 10 μF aluminum electrolytic capacitor (Mouser P/N 80-ESK106M016AC3AA)
- 1,000 μF aluminum electrolytic capacitor (Mouser P/N 647-UVR1C102MPD)

- Two 3.5 mm audio cables (Adafruit P/N 2698)
- Music player and speakers of your choice

HOW IT WORKS:
TURNING MUSIC INTO DATA

To visualize your music, the visualizer must analyze the content of an audio signal and display something in response on the LED strip. But how do we analyze an audio signal with Arduino code? The secret lies in the *Fast Fourier Transform (FFT)* algorithm, a mathematical technique dating back to the 19th century. We won't go into the math details because digital signal processing is quite complex, but the basic idea is that a signal that varies over time (like the voltage of an audio signal) can be analyzed and broken down into its frequency components. You're probably already familiar with this idea if you've seen a spectrum analyzer on stereo equipment and music players on your computer, shown in Figure 5-1.

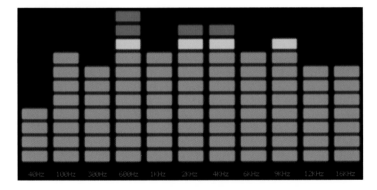

FIGURE 5-1:
A typical spectrum analyzer

The vertical bars represent frequency bands, with the bars on the left representing lower frequencies (bass) and those on the right representing higher frequencies (treble). The Arduino code in this project will sample the audio signal's voltage and perform the FFT algorithm on the audio samples to determine the signal strength in each frequency band. Then we'll use the levels of the low-frequency bands (the bass beat of the music) to create an interesting display on the LED strip.

In the visualizer circuit you'll plug your computer, phone, tablet, or other music device into a 3.5 mm (1/8 inch) input jack to take the music signal. You'll connect an output jack to your powered speakers, stereo amplifier, or whatever device you use to amplify your music and output it to speakers. In other words, the circuit sits between your music source and amplifier/speaker equipment so it can "listen" to the music and put on a show. The rest of the circuit consists of some simple components and an Arduino. There are some tricky aspects to dealing with audio signals in electronics, so this section describes a couple of the techniques we'll use to make the circuit work. This section will teach you quite a bit about audio processing!

Input Bias

An Arduino board can measure voltages on its six analog input pins, labeled A0 through A5 on the board. Your visualizer will connect the audio signal for your music to Arduino pin A0 to take a large number of sample measurements very quickly, then apply the FFT algorithm to the signal to transform it into data the Arduino can analyze.

If you're familiar with analog input measurement on Arduino, you know that a call to the `analogRead()` function, which reads data from the analog pins, returns a value in the range of 0–1023, which represents the measured voltage in the range of 0–5 V. But there is a problem when we want to measure an analog audio signal: audio is an *alternating current*, not direct current. That is, the voltage swings above ground (0 V) and then *below* ground to a negative voltage (see Figure 5-2).

FIGURE 5-2:

A simple audio signal oscillating between positive and negative voltages

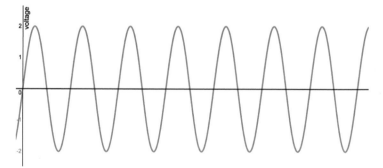

The Arduino cannot measure voltages below 0 V, so we cannot connect an audio signal to pin A0 without potentially damaging the Arduino. How do we solve this?

The solution is to "bias" the voltage up to a higher level so that it is not centered around 0 V, but around a higher voltage instead. That means when the voltage swings low it won't dip under 0 V. We accomplish this with a voltage divider made out of two resistors of equal value. The audio signal has one resistor connecting it to 5 V and another connecting it to 0 V (Figure 5-3). This biases the signal up to the midpoint between 0 and 5 V, or 2.5 V.

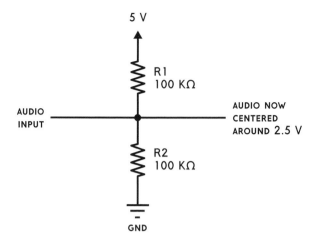

5 V

R1
100 KΩ

AUDIO
INPUT

AUDIO NOW
CENTERED
AROUND 2.5 V

R2
100 KΩ

GND

FIGURE 5-3:
An input bias circuit for
an audio signal

Figure 5-4 shows the same signal, but centered around 2.5 V instead of 0 V. Now the Arduino can measure audio voltages without worrying about them swinging below ground. Audio signals are usually only a few volts from peak to peak, so this approach works well.

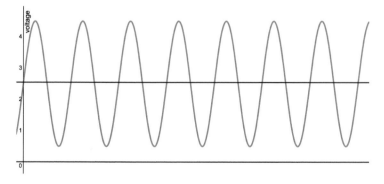

FIGURE 5-4:
An audio signal now biased
up 2.5 V

Sensitivity Control

The audio signal needs to be strong enough that the voltage swings over a large enough range for the visualizer to be able to effectively analyze it. But some audio sources don't output a really strong signal. We need to give our circuit the ability to adjust the sensitivity so that we can work with weak signals. This section describes a clever trick to accomplish this.

Recall from the previous section that the Arduino can measure voltages between 0 V and 5 V, and that we've biased the audio voltage up to be centered around 2.5 V. If the audio signal is weak and the voltage doesn't vary much, we'd have a signal like that shown in Figure 5-5. This might be the case in a "line-level" signal where the peak-to-peak voltage is only 1 V.

FIGURE 5-5:

A weak audio signal

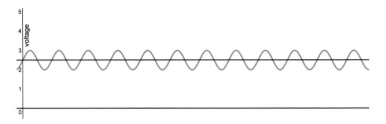

Measuring this signal will result in analogRead(0) values around 512 (the midpoint between 0–1023), but they won't vary enough to give us a good analysis.

Fortunately, the Arduino board has a special pin called the *analog reference (AREF)*. We can provide a voltage on this pin and tell the Arduino code that this is the highest voltage we are going to read on the analog input. For example, to measure voltages between 0 V and 2 V and have the analogRead() value use the full 0–1023 range, we provide 2 V on the AREF pin. A 2 V measurement will give the value of 1023 instead of something much lower.

In this circuit we'll use a potentiometer to provide a voltage to the AREF pin. For higher sensitivity, we simply provide a lower voltage by turning the potentiometer.

Easy, right? Not so fast! Remember that the audio signal is centered around 2.5 V. If we lower the analog reference too low, we'll be ignoring the upper part of the signal because the top peaks will be "cut off." In this case we will just get garbage from the FFT algorithm. The solution is that we *also* lower the top voltage on the input bias circuit so that the bias circuit will center the audio signal around the midpoint between 0 V and the voltage we provide to AREF. Figure 5-6 shows the situation where we have a weak audio signal that swings only about 1 V from peak to peak. The potentiometer in the circuit provides an AREF voltage of 2 V, which we also use as the top of the bias circuit. Now the audio signal is centered around 1 V and fills a much larger range of analogRead() values. In effect, we've amplified the signal! We didn't actually increase the voltage of the signal; we simply reduced the range of voltages that we're measuring so that the weak signal fills more of the range. Clever.

FIGURE 5-6:

A weak audio signal
with sensitivity adjusted
to set top voltage to 2 V

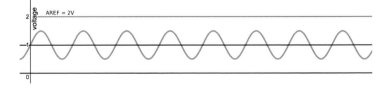

BUILD IT

It's time to start building the music visualizer circuit on a breadboard. These instructions will show this process in multiple steps to make it easier to build up the circuit. You may choose to lay out your components on the breadboard differently, but these diagrams are intended to give you the most clarity. Let's get started!

1. **Add the audio jacks to the breadboard.** Connect the left and right channels of the input jack directly to the left and right channels of the output jack, respectively, as shown in Figure 5-7. This means that the music flows through to output. The jack middle pins should connect to ground. Also connect the power rails of each side of the board with red and black wires as shown.

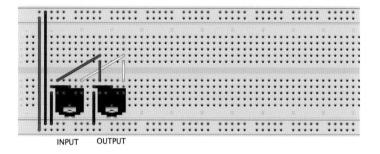

INPUT   OUTPUT

FIGURE 5-7:
Audio jacks and power rail connections

2. **Filter out noise with an audio summing circuit.** Now connect the left and right channels together through a resistor on each channel to prevent crosstalk (interference) between them. Add a 4.7 kΩ resistor on each channel, as shown in Figure 5-8. Add the 10 µF capacitor with the negative terminal (the shorter lead) on the left, connected to the output jack via the 4.7 kΩ resistor. Then add a 0.047 µF (or 47 nF) capacitor and connect one leg to the positive terminal of the 10 µF capacitor and the other leg to ground, as shown. These help to filter out noise and block DC current.

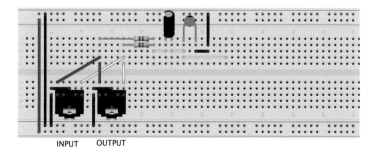

INPUT   OUTPUT

FIGURE 5-8:
Audio summing circuit and filter

3. **Build up the input bias and analog reference circuit.** Recall that our voltage divider is made of two 100 kΩ resistors and will center the audio signal between 0 V and the analog reference voltage, which we control with the 10 kΩ potentiometer. Insert those resistors, with one leg of the first resistor placed directly in the ground rail, and the other leg connected to the second 100 kΩ resistor, as shown in Figure 5-9. Make sure the second resistor is inserted adjacent to the first, as shown. Now place a potentiometer in the breadboard and connect the rightmost pin to ground through a 2.2 kΩ resistor. Connect a 4.7 kΩ resistor to the middle pin of the potentiometer—this will eventually connect to the AREF pin on the Arduino—and make sure this resistor straddles the center divide on the breadboard. This middle pin should also connect to the second leg of the second 100 kΩ resistor. The final empty pin of the potentiometer connects to the power rail.

FIGURE 5-9:

Input bias components

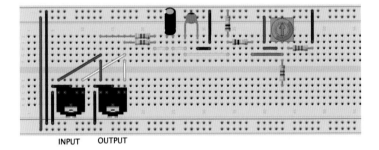

INPUT   OUTPUT

4. **Connect a power jack to power the circuit.** Connect the power jack to the power rails of the breadboard. Also connect the large 1,000 µF capacitor to the power lines as shown in Figure 5-10. Electrolytic capacitors are polarized, so pay attention to the polarity: the positive lead is longer than the negative lead and should connect to the positive rail. There is also a white stripe on the negative side of the capacitor. This capacitor provides a reservoir of voltage in case the LED strip draws a lot of current all at once. For example, setting all LEDs on the strip to white at the same time would draw a lot of current from the circuit, but the capacitor would help smooth out any resulting voltage dips so the circuit doesn't overload.

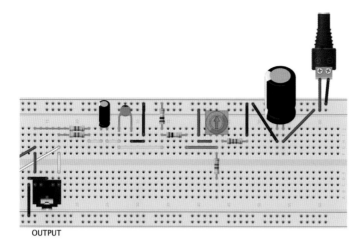

OUTPUT

FIGURE 5-10:
Power connection for the
LED strip

5. **Connect the LED strip to your circuit.** The LED strip has
   three connections: 5 V, ground, and data in (DIN). These electrical connections are not always in the same order on all LED
   strips. For example, the DIN connection is often in the middle,
   but Figure 5-11 shows it at the top. Just pay close attention to
   the labels on your LED strip. You may need to solder wires to
   your LED strip if yours doesn't have a connector of some kind
   (turn to the appendix for instructions on soldering). Whatever the
   case, connect the 5 V, ground, and DIN connections to the circuit
   as shown in Figure 5-11, with the DIN pin connected to a 470 Ω
   resistor (the other end of the resistor will connect to the Arduino).

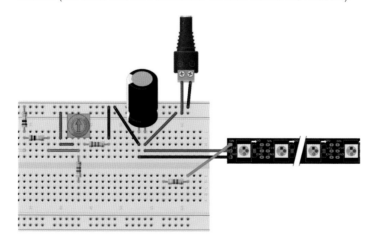

FIGURE 5-11:
Connecting the LED strip

6. **Wire up the Arduino.** We're almost done! The last wires connect to your Arduino board, as shown in Figure 5-12. The green
   wire connects Arduino pin 6 to the 470 Ω resistor on the DIN

connection. The blue wire connects to the Arduino AREF pin. The yellow wire is the audio signal and connects to Arduino analog pin A0. The last connection you need to make is wiring the breadboard ground signal to an Arduino ground pin marked GND, but don't make this connection just yet. If there is a program already running on your Arduino, it could damage parts of your circuit when you connect your circuit to the Arduino. So, to be safe, we'll load the sketch for this project first and then finish connecting the Arduino.

FIGURE 5-12:

Connecting the Arduino
to the circuit

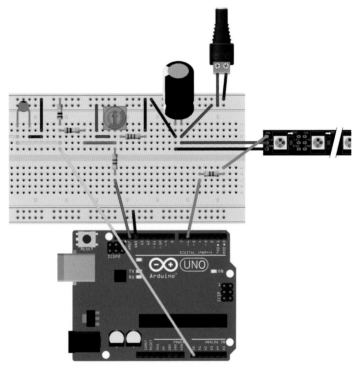

CODE IT

This circuit won't do anything without some awesome code on the Arduino! The Arduino sketch we'll use is fairly complex and there isn't room to print it all here, but this section will describe how it generally works so you can learn the basics. Download the sketch from the book's resources at *https://nostarch.com/LEDHandbook/*.

How It Works

The Arduino program's job is to "listen" to the music by capturing voltage samples in a buffer and then perform an FFT analysis on those samples to determine the levels of each frequency band. This gives us data similar to an ordinary spectrum analyzer—the signal strength of each frequency band at any particular moment.

Instead of just displaying columns of LEDs, though, we do something much more interesting. Every time a new maximum or peak in a frequency band is detected, the code generates a pair of moving pixels that originate in the middle of the LED strip and move outward toward the ends. The speed of the moving pixels depends on the peak's strength, and the pixels fade in brightness as they move outward. The colors of the peaks vary randomly. Note that we pay attention only to the bottom three frequency bands (out of eight bands) so that we are visualizing the music's beat.

Here's the code for the main loop, with explanatory comments:

```
void loop() {
  // While the ADC interrupt is enabled, wait. The program is
  // still gathering audio samples in the capture buffer.
  while (ADCSRA & _BV(ADIE)) {
    // Wait...
  }

  // The sampling interrupt has finished filling the buffer,
  // so show the output we computed last time through the
  // loop. This sends the data to the LED strip.
  strip.show();

  // Perform the FFT algorithm to convert samples to
  // complex numbers.
  fft_input(capture, bfly_buff);

  // Now that we've updated the LED strip and processed the
  // audio samples with FFT, we can resume the collection
  // of audio samples in the sample buffer. The interrupt
  // service routine (ISR) will run while we compute the next
  // LED output based on the audio samples we captured.
  startSampling();

  // Perform the rest of the FFT computation:
  fft_execute(bfly_buff);          // Process complex data
  fft_output(bfly_buff, spectrum); // Complex -> spectrum

  // Now call this to analyze the audio. See comments in this
  // function for details.
  analyzeAudioSamples();

  // The peak values for each of the 8 bands has been
  // computed. A bit in the 8-bit value newPeakFlags
```

```
    // indicates whether the analysis found a *new* peak
    // in the band.
    for (i = 0; i <= CUTOFF_BAND; i++) {
      // If a new peak was found in band i...
      if (newPeakFlags & (1 << i)) {
        // Map the peak value to a magnitude in range [0,255].
        // We pass in the band number because the mapping is
        // different for different bands.
        uint8_t magnitude = getMagnitude(i, bandPeakLevel[i]);

        // A nonzero magnitude means that the peak value is
        // large enough to do something visually with it. We
        // ignore small peaks.
        if (magnitude > 0) {
          // We want to store the information about the peak
          // in a peak_t structure. When we actually draw a
          // visualization, the peak_t structures in peaks[]
          // represent the "visually active" band peaks.

          // Look through the list of peak structures 'peaks'
          // for an unused one.
          for (j = 0; j < N_PEAKS; j++) {
            if (peaks[j].magnitude == 0) {
              // Unused peak found.
              peakIndex = j;
              break;
            }
          }
          // If an unused one not found, we use the last one
          // that was used (peakIndex).

          // Initialize the structure.
          peaks[peakIndex].age = 0;
          // A random component for a visualization to use. For
          // example, to shift the color a small amount.
          peaks[peakIndex].rnd = random(255);
          peaks[peakIndex].baseColor =
            getRandomBaseColor(peaks[peakIndex].rnd);
          peaks[peakIndex].magnitude = magnitude;
        }
      }
    }

  // Clear the last frame.
  strip.clear();
  // Draw the peaks on the LED strip.
  doVisualization();
} // end loop()
```

There's a lot more to the code than this main loop, and most of the heavy lifting is performed in the `analyzeAudioSamples()` and `doVisualization()` functions. You certainly don't have to understand how all of it works to enjoy the project, though. In the next section you will load the code onto your Arduino.

Get the Code

The full source code for the project is in the book's resources at *https://nostarch.com/LEDHandbook/*. You'll also find the library used to drive the RGB (red-green-blue) LED strip in the resources, which you'll need to install into the Arduino IDE. Adafruit has an excellent guide on installing Arduino libraries at *https://learn.adafruit.com/adafruit-neopixel-uberguide/arduino-library-installation*. After installation, your Arduino directory for this code should have this structure:

```
your_sketchbook
  |
 +--libraries
  |    |
  |    +--ffft
  |    |
  |    +--Adafruit_NeoPixel
  |
 +--RealtimeMusicVisualizer
       |
       +--RealtimeMusicVisualizer.h
       +--RealtimeMusicVisualizer.ino
```

Now open the Arduino IDE and load the RealtimeMusicVisualizer sketch. Compile and upload it to your Arduino. It is usually a good idea to disconnect the Arduino from your new circuit before powering on the Arduino. After you've uploaded the code onto the Arduino, you can make the four wire connections to your circuit described earlier.

USE IT

Using the music visualizer is as easy as making a few connections:

1. Connect the audio input jack to your music source with a 3.5 mm audio cable.

2. Connect the audio output jack to your stereo receiver or some powered speakers.

3. Connect your 5 V power supply to the power jack on the breadboard circuit. The power supply must be capable of supplying at least 2 A of current.

4. Connect power to your Arduino board, either with a USB cable or other power adapter. Otherwise, you can provide power to the Arduino from the 5 V breadboard supply by connecting a wire to the 5 V pin on the Arduino.

Now turn on the music and enjoy the show! You'll want your music player to be at a fairly high volume, especially if it's a small device like a phone or tablet. Computers tend to output a stronger signal. If the music is not producing much of a visualization, increase the sensitivity of the potentiometer on the breadboard by turning the knob clockwise until you can clearly see the beat of the music.

TAKE IT FURTHER

If you're experienced with Arduino programming, you'll probably want to play with the code and make it do different things. Here are some ideas to explore:

- **Change the number of LEDs.** If your LED strip has a different number of LEDs, you should adjust the code. For example, if you have a 3 m strip with 180 LEDs, change the line in the file *RealtimeMusicVisualizer.ino* that defines N_LEDS to the new value:

```
#define N_LEDS 180
```

- **Make the pixels move in one direction.** Instead of having the moving pixels move from the center of the strip toward both ends, you can make them start from one end only. Set SINGLE_DIRECTION to true in *RealtimeMusicVisualizer.ino*:

```
#define SINGLE_DIRECTION true
```

- **Make the code respond to more frequency bands.** The code pays attention only to the bottom three frequency bands. By changing the value of CUTOFF_BAND, you can listen to more bands. The current value is 2 because the bottom three bands are numbered 0, 1, and 2. If you wanted to respond to the bottom five bands, you'd change the line to this:

```
#define CUTOFF_BAND 4
```

- **Change the brightness.** Is the visualizer too bright for you? Maybe you aren't throwing a huge party but just adding some bling to your dorm room. You can turn down the brightness by reducing the value of MAX_BRIGHTNESS. The default is 255, so try a lower value like 100:

```
#define MAX_BRIGHTNESS 100
```

- **Change the way the colors vary over time.** The visualizer displays pixels of two colors that change over time. For each of these two colors, the speed at which the color changes over time is controlled by the values MIN_COLOR_CHANGE_COUNT and MAX_COLOR_CHANGE_COUNT. When a new color is chosen, a value between these min and max values is set, indicating how many pixels will be that color until a new color is chosen. The default min and max values are 5 and 10. If you want the chosen colors to appear for longer before they change, increase these values. If you want every pixel to be a new color, you can set both values to 1:

```
#define MIN_COLOR_CHANGE_COUNT 1
#define MAX_COLOR_CHANGE_COUNT 1
```

- **Write your own visualization.** If you have an idea for a whole new visualization, you can modify or rewrite the function doVisualization(). This function draws the moving pixels that are defined by the peak_t structures in the array peaks. If you study the code for a while, you will understand how it works. This project is based on a more complex visualizer that has many different visualizations: the Lumazoid board (a nootropic design product). You can get some ideas for different visualizations from the Lumazoid source code at *https://github.com/nootropicdesign/lumazoid*.

SUMMARY

If you've built the circuit and got everything working, congratulations! You now have a very cool way to light up your music, and hopefully you learned some things along the way, too. If you read all the material in this chapter about how music is analyzed using digital signal processing and how to deal with audio signal voltages using simple electronics tricks, you already know more than most people about the true nature of audio signals.

AUTOMATED INFRARED REMOTE CONTROL

BY KAAS BAICHTAL

IN THIS PROJECT, YOU'LL USE INFRARED LEDS AND A RASPBERRY PI TO AUTOMATI-CALLY OPERATE A REMOTE CONTROLLED DEVICE.

HOW IT WORKS: INFRARED LIGHT

In the early 1800s, astronomer William Herschel discovered that sunlight split into colors by a prism is noticeably warmer the closer it is to the red end of the spectrum. However, the area of the spectrum beyond red is even warmer. It turns out there is another, warmer color of light—called *infrared (IR)*—which humans can feel but not see.

Infrared means "below red," so named because it has a lower *frequency* (waves per second) than visible red light. The electromagnetic spectrum chart in Figure 6-1 shows the relationships between all radiant energies, including light, radio waves, microwaves, gamma rays, and X-rays.

FIGURE 6-1:

The electromagnetic spectrum

← LONGER   SHORTER →

| WAVELENGTH (IN METERS) | 10^3 10^2 10^1 1 10^{-1} 10^{-2} 10^{-3} 10^{-4} 10^{-5} 10^{-6} 10^{-7} 10^{-8} 10^{-9} 10^{-10} 10^{-11} 10^{-12} |

COMMON NAME OF WAVE

RADIO WAVES • VISIBLE • INFRARED | ULTRAVIOLET • "HARD" X-RAYS

MICROWAVES • "SOFT" X-RAYS • GAMMA RAYS

| FREQUENCY (WAVES PER SECOND) | 10^6 10^7 10^8 10^9 10^{10} 10^{11} 10^{12} 10^{13} 10^{14} 10^{15} 10^{16} 10^{17} 10^{18} 10^{19} 10^{20} |

Infrared LEDs

Infrared LEDs have been around since the early 1960s. In fact, they were invented a year or two before LEDs that emit visible light. Figure 6-2 shows an example of an infrared LED.

FIGURE 6-2:

An infrared LED. The larger section visible inside corresponds to the negative lead.

Like all diodes, an LED is a semiconductor. Current passes through a diode more easily in the forward direction than it would in reverse. It also emits a small quantity of electromagnetic radiation as it moves forward through the diode. Where this emission lies on the electromagnetic spectrum depends upon what materials were used to manufacture the diode. Regular diodes made of silicon and germanium do not tend to emit light, while other semiconductor materials can emit various colors of visible light, as well as ultraviolet and infrared. Infrared LEDs are usually made with gallium arsenide (GaAs) or aluminum gallium arsenide (AlGaAs).

When an LED is reverse-biased (hooked up backward), no light will be produced.

Line-of-Sight Communications

Infrared LEDs have dozens of important applications, including data communications. Circuits that send data using IR LEDs are called *transmitters*. Circuits that receive the infrared light and convert it back to data are called *receivers*.

Because infrared components transmit and receive light, they need to be able to "see" each other to successfully communicate (see Figure 6-3). This is commonly referred to as *line-of-sight communications*. Nothing opaque can be in the way, and the narrower the light beam the transmitting LED emits, the more precisely the transmitter needs to be aimed at its receiver. There are ways to steer or reflect light in non-line-of-sight directions when necessary, however—for example, through the use of mirrors, fiber optics, and light pipes.

FIGURE 6-3:
This IR remote has a hole for its transmitter's LED to beam out, while the air conditioner it controls has a small red window to allow IR light to shine in upon its receiver.

While it may seem like a disadvantage, the line-of-sight nature of infrared communications actually has its pluses: the IR transmissions are unlikely to interfere with other IR communications happening in completely different rooms or buildings. Anybody who has ever had their ceiling fans turned on and off by a neighbor's radio remote will be able to appreciate that!

For this project we will use simple line-of-sight communications, orienting the IR transmitter attached to our Raspberry Pi so that it can be seen by the air conditioner without problems.

Data Transmission

In the simplest possible sense, an IR transmitter communicates data by turning its LED on and off. For example, it would be possible to communicate using Morse code with an LED. However, several complicated languages and systems, called *protocols*, have been developed specifically for data communication over IR. Over the decades, more and more information has needed to be conveyed through the messaging sent by remotes. This has sparked many inventive ways of increasing the signal-to-noise ratio, error correction, and data carrying capacity. Here's an overview of some of the things that might be included in a protocol specification:

- The frequency of infrared light used
- The carrier frequency or frequencies used
- Number of devices supported
- Number of commands per device
- Number of bits used in encoding
- The encoding procedure
- The key-to-code mapping
- The data rate, which may be variable
- Modulation schemes of the transmitted pulses and/or the spaces between them

Before the advent of programmable microcontrollers, transmitters and receivers needed additional circuitry to accomplish everything a protocol required. A general lack of compatibility led to the development of "universal remotes," which either have the payloads of many other remotes preprogrammed into them, or simply record the signals of other remotes and parrot them back upon command. This latter is the procedure we'll be using in this project to begin "teaching" the Raspberry Pi how to control the A/C, using a USB IR Toy (see Figure 6-4). The USB IR Toy has a transmitter and receiver on the same circuit board, so we can use it to record the signals from the air conditioner's remote as well as mimic them.

FIGURE 6-4:

The USB IR Toy

GET THE PARTS

The instructions in this project will use the equipment listed here. New to electronics projects? Stick with this parts list and I'll make it easy!

If you have more experience and are able to modify the steps, feel free to substitute your own transmitter, receiver, computer, OS, and A/C or other item to be controlled.

- USB IR transmitter and receiver (I used a Dangerous Proto-types USB IR Toy 2, which is available either prebuilt or in kit form. The prebuilt ones are available at Seeed Studios with SKU #102990037; *https://www.seeedstudio.com/USB-Infrared-Toy-v2-p-831.html*.)

- USB cable with A and mini-B connectors (for example, Adafruit P/N 260)

- Raspberry Pi (Any model will work, but it will need to have Raspbian installed via NOOBS—see "Getting Started with the Raspberry Pi" on page 13 for instructions on setting that up.)

- An air conditioner that has an IR remote (I used a Haier HWR08XC7.)

The rest of this section deals with modifications, so if you're using the parts listed, feel free to skip to "Build It" on page 117.

Any computer that has a USB port for the IR Toy 2 and is capable of running LIRC (Linux Infrared Remote Control) will work for this project. I chose a Raspberry Pi because I could get one bundled with a case, power supply, and NOOBS SD card cheaply on Amazon. Raspbian is the distribution being used for other projects in this book, so that's what I'll be providing instructions for, but everything we do here should be modifiable to work on your machine.

Whatever you use, you'll need the LIRC package (*http://www.lirc.org/*) for this project. LIRC is handy for this kind of task because 1) it can be installed on many flavors of Unix-like and Windows OSes; 2) it allows multiple programs to listen to one receiver; 3) it comes with helpful tools for memorizing remote controls and sending commands manually; and 4) its website has plenty of hardware compatibility info, so you can easily adapt this project to work with other hardware.

For example, there are lots of preconfigured *.conf* files for different handheld remotes, located at *http://lirc-remotes.sourceforge.net/remotes-table.html*. In theory, you can control any device that works with one of those listed IR remotes without doing any coding or even having to scrabble in your couch cushions to find the remote. But few of those preconfigured remotes are intended for air conditioners, so for this project we'll just create our own *.conf* file and not stress about finding a matching one.

In fact, since we'll be creating our own *.conf* file anyway, you could use any IR-controllable device and the procedure will be exactly the same, as long as you still have its physical remote to imitate the codes from. Xbox? Roomba? DVD player? No problem!

If you intend to buy an A/C unit for the project, just go for one that, besides having the IR capability, meets your cooling needs and is a common everyday brand where you live.

For my IR transmitter and receiver, I chose the USB IR Toy 2 by Dangerous Prototypes for three reasons: it's fully compatible with LIRC; it's relatively inexpensive at $20 for a preassembled version (or almost nothing if you want to assemble it yourself); and finally, since it has both a transmitter and a receiver, I only have to deal with one device. Although actually controlling the air conditioner requires just a transmitter, we need the receiver as well since we're creating our own initial configuration files.

BUILD IT

Let's make the project! The following steps assume you already have an assembled IR transmitter and receiver, A/C unit and remote, and sudo root privileges on your Pi. (Raspbian provides the "pi" user for this purpose.)

1. **Upgrade your USB IR Toy 2's firmware.** Follow the directions on the Dangerous Prototypes website's USB IR Toy 2 page (*http://dangerousprototypes.com/docs/USB_IR_Toy_firmware _update*) to upgrade its firmware to the latest version, which was officially v22 at the time of this writing. Performing a firmware upgrade could save you a lot of trouble depending on which revision your unit shipped with, as some earlier versions did not support transmitting!

   You can upgrade your firmware via the USB port using almost any computer. After downloading and unzipping the firmware upgrade package from the Dangerous Prototypes website, follow the instructions provided on the site. The procedure and results will vary depending on your setup, so I won't go into a lot of detail except to say that on my Windows 7 desktop I had to manually install the driver included in the *inf-driver* subdirectory, a detail not mentioned in the instructions.

   Once the computer recognizes the USB IR Toy 2, run the *update-USBIRToy.v22.bat* file in the firmware subdirectory and, when it asks, give it the port number assigned to your device (see Note for details).

   I found it helpful to pause momentarily every time I plugged in or changed anything, as Windows re-recognized the USB IR Toy 2 more than once during this process, and the interruption caused the firmware updater to not find the device. Once the device was successfully recognized, the firmware upgrade process was quick and easy.

2. **Upgrade the Raspberry Pi's OS and firmware.** I recommend making sure your operating system is 100% patched/upgraded and the Raspberry Pi has also had its firmware fully updated to the latest version before you begin.

NOTE

At the time of this writing, there were also user-contributed versions v23 and v24, which could be found in the forums on the website.

NOTE

If you did not receive a COM port, and can't find it listed as a COM port in your Device Manager or equivalent, just press ENTER when you get to this question. You only need the port number if you were automatically assigned one.

3. **Check LIRC's dependencies.** Now make sure your system has everything LIRC requires. The list can be found at *http://www.lirc.org/html/install.html#dependencies*. I checked by entering the following lines on the Linux command line:

```
$ which make
$ which gcc
$ which g++
$ which ld
$ which modinfo
$ which pkg-config
$ which xsltproc
```

Each command should produce the full path of the item in question. If there is no response, the item is missing from the system. Two more checks you should perform are:

```
$ dpkg -l python3-yaml
$ dpkg -l raspberrypi-kernel-headers
```

If you have these, you should see some information about the package in question; if you don't, you should see a complaint saying the packages were not found. My fresh NOOBS install of Raspbian lacked both of these latter packages as well as xsltproc from the preceding list, so I installed the necessary items as follows:

```
$ sudo apt-get install xsltproc python3-yaml \
raspberrypi-kernel-headers
```

4. **Install LIRC.** Next, download the LIRC source, which can be obtained on SourceForge at *https://sourceforge.net/projects/lirc/files/LIRC/*. Choose the directory with the most recent version of the code, save the *.tar.gz* file for that version, and then go to the save location. To extract and install LIRC, enter the following on the Linux command line, inserting your correct version number:

```
$ tar xjf lirc-version.tar.bz2

$ cd lirc-version
$ ./configure
$ make
$ sudo make install
$ sudo ldconfig
```

5. **Connect the USB IR Toy 2.** Now, power down your Pi, then plug in the USB IR Toy 2. Make sure to power the Pi down every time you plug and unplug your Toy, or you risk blowing stuff up! After rebooting your machine, look in the dmesg logfile to see if the Pi found your device:

```
$ dmesg | grep -e usb
```

You should see output like this:

```
[    2.982746] usb 1-1.2: new full-speed USB device
number 4 using dwc_otg
[    3.090984] usb 1-1.2: New USB device found,
idVendor=04d8, idProduct=fd08
[    3.092965] usb 1-1.2: New USB device strings: Mfr=1,
Product=2, SerialNumber=3
[    3.096209] usb 1-1.2: Product: CDC Test
[    3.097827] usb 1-1.2: Manufacturer: Dangerous
Prototypes
[    3.099458] usb 1-1.2: SerialNumber: 00000001
[    4.358474] usbcore: registered new interface driver
cdc_acm
```

This output shows that the Pi found my USB IR Toy 2 and was able to automatically find and load the correct kernel-level driver for it, cdc_acm. If your machine doesn't load a kernel-level driver after finding the Toy, you can modify your system to load it each time the machine is booted. If you're using Raspbian, you can do this by editing the file *etc/modules* and adding the following line:

```
cdc_acm
```

Once that's working, you can use dmesg as follows to find the device name that's been assigned to the USB IR Toy 2 (which you'll need in the next step):

```
$ dmesg | grep -e cdc_acm
```

My output gave me the following:

```
[    4.332559] cdc_acm 1-1.2:1.0: ttyACM0: USB ACM device
[    4.358474] usbcore: registered new interface driver
cdc_acm
[    4.360314] cdc_acm: USB Abstract Control Model driver
for USB modems and ISDN adapters
```

NOTE

If you're using a completely different IR transmitter/ receiver for your project, you'll quite likely have a different kernel-level driver. That too can be loaded in /etc/modules.

In this case, my USB IR Toy 2 has been assigned to device */dev/ttyACM0*. Write down your device assignment.

6. **Configure *lirc_options.conf*.** LIRC has put some default configuration files in */usr/local/etc/lirc*. For this step you need to edit */usr/local/etc/lirc/lirc_options.conf* as superuser and change the `driver` parameter (not the kernel-level driver in this case, but the LIRC plug-in) to `irtoy` and the `device` parameter to the full path of the one you saw in your `dmesg`. Open your */usr/local/etc/lirc/lirc_options.conf* file and change those parameters to match the following:

```
driver = irtoy
device = /dev/ttyACM0
```

If you are using something besides a USB IR Toy 2, you can find a list of all other available plug-ins using the following command:

```
$ lirc-lsplugins
```

While you're in there, you may also need to change the default locations for `pidfile` and `output` so they match where they were actually installed on your system. In my case, the NOOBS-installed Raspbian, I had to prepend */usr/local* to each one. You'll know you failed if LIRC complains it can't write the PID file when you try to run it!

```
output = /usr/local/var/run/lirc/lircd
pidfile = /usr/local/var/run/lirc/lircd.pid
```

7. **Paranoia check.** To make sure you're on track, check that you used the right device and driver with the following command:

```
$ sudo mode2 -H irtoy -d /dev/ttyACM0
```

Point any IR remote at the USB IR Toy 2 and push some buttons. You should see results flow across the screen, indicating that serial data is available on the Pi as a result of your button pushing. Mine started out like this:

```
Using driver irtoy on device /dev/ttyACM0
Trying device: /dev/ttyACM0
Using device: /dev/ttyACM0
```

```
Running as regular user pi
space 1000000
pulse 3861
space 1000000
space 1000000
pulse 21
space 76799
pulse 127
space 12501
pulse 63
space 1112467
pulse 42
space 1000000
space 1000000
pulse 5119
space 4543
pulse 554
space 575
pulse 533
space 575
pulse 554
--snip--
```

When you're done, press CTRL-C to exit that program.

8. **Pick names for your buttons.** The information on each remote control your LIRC knows about will be stored in the *.conf* files located in */usr/local/etc/lirc/lircd.conf.d*. The next step is to create a file there for the air conditioner. We'll do this using a program called *irrecord*.

Each button on your remote will need a name in the configuration file. Write your choices down so they'll be handy when you go to record your button presses. LIRC has some standardized button names that you can either use or disable; for simplicity we'll use these standard names in this project. You can get the list by entering:

```
$ irrecord -l
```

For my buttons I chose and wrote down the following:

```
on/off: KEY_POWER
mode: KEY_MODE
speed: KEY_FASTFORWARD
timer: KEY_TIME
temp/time up arrow: KEY_UP
temp/time down arrow: KEY_DOWN
```

9. **Extra credit: get the model number of your remote.** While not strictly necessary for this project, it's a nice touch (and LIRC convention) to have your config file named after the remote you are using. I was able to find the model number of mine, a Haier AC-5620-30, by scrutinizing the spare parts section on Haier's website and finding a match.

NOTE

If at any time during the recording process the USB IR Toy 2 stops functioning, I would recommend power-cycling your Pi and unplugging/replugging the USB IR Toy 2's cable while the power is off. I don't know where the fault lies, but recording can be a bit dodgy!

10. **Record the codes sent by the remote.** Now you'll create a crude initial configuration file for your air conditioner's remote by naming the buttons and pushing them such that the raw (not decoded yet) code is recorded and associated with its correct button name. Recording the raw signal will allow LIRC to reproduce it as an output later when you program your computer to control the A/C.

    First move to a scratch directory where you have write privileges:

```
$ cd /home/pi/
```

With your remote in hand and the USB IR Toy 2 arranged so the LEDs point toward you, enter:

```
$ sudo irrecord -H irtoy -d /dev/ttyACM0 -f
```

Make sure to use your correct driver and device names. The program will step you through some processes to gather technical information on the signals your particular remote sends. First you'll be asked to enter the name of your remote. Then it will ask you to push many random buttons, which will allow it to see how buttons generally work in this new and unfamiliar remote's unknown protocol. Once it is confident it can tell one button push from another, it will ask you to enter your chosen button names one by one, and demonstrate the button push for each entry.

Follow all the instructions through to the end. When it finishes, the contents will be saved to a file that starts with the remote model name you provided, and ends in *.lircd.conf*.

If you later find that some button(s) did not record properly, you can re-record just those ones by specifying the config file to be updated using -u. For example:

```
$ irrecord -H irtoy -d /dev/ttyACM0 -f -u \
Haier_AC-5620-30.lircd.conf
```

This technique can also be used to adapt someone else's config file for your own remote, if they are compatible enough.

11. **Edit your *.lirc.conf* file.** Using your favorite text editor, open your newly created config file, and you should see that it has three sections:

- First comes a comments area, surrounded by hash marks (#), with annotated blanks for you to fill in your remote and other relevant details. The LIRC community likes this information to be in this uniform format and fully fleshed out in the event that you decide to share the file with others. If you intend to keep the file to yourself, however, filling in the comments is optional.

- Next, directly under `begin remote`, there should be a section with your remote's name and some basic serial communications characteristics of its signal. This information was gathered by irrecord during the process you went through in step 10, and will be different for different kinds of remotes. Mine looked like this:

```
begin remote
    name    Remote_Haier_AC-5620-30.conf
    flags RAW_CODES|CONST_LENGTH
    eps          30
    aeps        100
    gap        108529
```

- Third, there will be a list of the button names you chose, each accompanied by a block of code that irrecord has associated with that button. You may notice that some of them look suspiciously short, or otherwise dramatically different from their fellows. These may well turn out to be misrecorded ones. You'll find out for sure in the next step. Here's my first button from that section:

```
begin raw_codes

    name KEY_POWER
        8981    4565    533    575    533    597
         511     575    533   1706    511   1727
         533     575    511    597    533   1706
         511    1706    554   1685    533   1706
         533    1706    533    575    511   1727
         533    1706    533    575    533   1706
         533     575    511    597    511   1727
```

533	1706	533	575	533	575
533	597	511	575	533	1706
533	1706	533	597	511	575
533	1706	533	1727	511	1706
533					

12. **Test your .lirc.conf file.** Move a copy of your .lirc.conf file to the configuration directory as follows:

```
$ sudo cp Haier_AC-5620-30.lircd.conf \
/usr/local/etc/lirc/lircd.conf.d/
```

If you're not using the default config file, *devinput.lircd.conf*, for anything, you may wish to change its name to something like *devinput.lircd.conf.unused* so LIRC doesn't try to use it.

Next, let's test the config file. Start LIRC by entering:

```
$ sudo lircd
```

With LIRC running, test your new config file by entering:

```
$ sudo irw
```

Now push the buttons on your remote while pointing at the USB IR Toy 2. Test each button and check that the name you chose for it pops up on the screen when you push it. Make a note if any button press comes up with the wrong name, or if any buttons cause the system to get confused (you'll know because it won't say anything and there will be a long pause before any more pushes work). Those are probably misrecorded.

When you are finished testing, press CTRL-C to get out of irw, then kill the LIRC process using:

```
$ sudo killall lircd
```

Take your list of erroneous buttons and use the –u technique described at the end of step 10 to re-record them. If irrecord won't connect to the USB IR Toy 2, you may have to power-cycle the Pi and USB IR Toy 2.

When you are done re-recording, run **sudo lircd** again. Then run **sudo irw** again and test the buttons. Repeat this step until your config file is 100% accurate! This is the file LIRC will look at when you ask it to generate codes to control your air conditioner with the USB IR Toy 2, so it pays to get this part right.

13. **Test control of the A/C.** Next you must verify that the recorded commands are accurate enough to control the air conditioner itself when reproduced by LIRC. You can do this using the `irsend` command. First, start LIRC:

```
$ sudo lircd
```

Now enter:

```
$ irsend list "" ""
```

You should see a list of any remote names in your config file(s)—in my case, just the one:

```
Haier_AC-5620-30
```

You can list the commands available for a given remote by inserting its name exactly as shown in the list:

```
$ irsend list Haier_AC-5620-30 ""
```

For me, this produces the following:

```
0000000000000001 KEY_POWER
0000000000000002 KEY_FASTFORWARD
0000000000000003 KEY_UP
0000000000000004 KEY_DOWN
0000000000000005 KEY_TIME
0000000000000006 KEY_MODE
```

The similarity with my list of chosen button names from step 8 is no coincidence! It is reading the names from the config file we created. It should now be possible to take one of those button names and combine it with the remote name to send a command like this:

```
$ irsend SEND_ONCE Haier_AC-5620-30 KEY_POWER
```

Boom! My air conditioner comes on. If you're going to be using this to control something long-term, you'll want to test all your buttons vigorously to make sure your LIRC and USB IR Toy 2 work together stably.

PATCHING AN UNSTABLE LIRC

The first time I reached this step with LIRC 0.9.4d, I found it very unstable. Both LIRC and the USB IR Toy 2 would die with an error message every several commands. If I replugged the USB IR Toy 2 and reran LIRC it would work again—for several commands. If you encounter this problem like I did, there is a patch available at *https://sourceforge.net/u/bengtmartensson/lirc/ci/experimental -irtoy/* that should fix it. Although diff patches are designed to be applied automatically, you'll likely have to apply this one manually due to differences in versions.

To apply the patch, first back up your *lirc_options.conf* file as follows:

```
$ sudo mv /usr/local/etc/lirc/lirc_options.conf \
/usr/local/etc/lirc/lirc_options.conf.backup
```

Then go back to the directory where you compiled LIRC, *~/lirc-0.9.4d/* in my case. From there, edit *./plugins/irtoy.c* and make the additions and subtractions specified in the green and red lines of the diff. Then clean up the directory and recompile like this:

```
$ ./configure
$ make clean
$ make
$ sudo make install
```

When done recompiling, reinstate your own *lirc_options.conf* file:

```
$ sudo mv /usr/local/etc/lirc/lirc_options.conf.backup \
/usr/local/etc/lirc/lirc_options.conf
```

If you changed the name of *devinput.lircd.conf* before, you'll have to delete the new copy the compilation added:

```
$ sudo rm /usr/local/etc/lirc/lircd.conf.d/\
devinput.lircd.conf
```

Now repeat step 13 and see if the system now performs reliably. The patch cleared up the problem completely for me!

14. **Analyze your .conf file.** When you are 100% satisfied with the performance of your .*conf* file, you can tell LIRC to analyze the remote's protocol much more thoroughly by using the following commands:

```
$ cd ~
$ sudo cp /usr/local/etc/lirc/lircd.conf.d/\
Haier_AC-5620-30.lircd.conf .
$ irrecord -a ~/Haier_AC-5620-30.lircd.conf
```

Technically you could have tried this immediately after recording the first time, but personally I found it worked much better if any errors or missing buttons were taken care of first. Once analyzed, my entire file had shrunk considerably and now looked like this:

```
begin remote

  name   Haier_AC-5620-30
  bits            32
  flags SPACE_ENC|CONST_LENGTH
  eps            30
  aeps          100

  header       8984  4554
  one           525  1712
  zero          525   584
  ptrail        529
  gap         108529
  toggle_bit_mask 0x0
  frequency     38000

      begin codes
          KEY_POWER              0x19F69867
          KEY_MODE               0x19F610EF
          KEY_FASTFORWARD        0x19F620DF
          KEY_TIME               0x19F658A7
          KEY_UP                 0x19F6A05F
          KEY_DOWN               0x19F6906F
      end codes

end remote
```

Not only is this much more compact and easier to read than the raw version shown (partially) in step 11, but much more is now known about the protocol used by the A/C. In addition to the header, frequency, and so on, the actual commands have been decoded, and are expressed in hex instead of as blinks of various lengths to be imitated. This means that LIRC actually knows what it is doing now, instead of parroting stuff it doesn't understand. That can only be a good thing when it comes to accurately controlling the A/C!

Cut and paste the analysis to replace the raw content in your original *.lircd.conf* file and kill and then restart LIRC to give it a try. It should work brilliantly.

NOTE

If nothing changed or happened when you tried to analyze your file, LIRC wasn't able to figure out your protocol. Just keep using your raw .conf file in that case.

15. **Run LIRC automatically.** Once you have LIRC configured properly, you can run it one last time and leave it running. You might also want to set it up to start automatically when you start your Pi, if you plan to set up tasks for your air conditioner controller that need to keep working if the machine gets rebooted. To do that on a Raspbian system, edit your */etc/rc.local* file as superuser and add the line:

```
lircd
```

16. **Program the Pi to run your A/C.** Now that you have your USB IR Toy 2 talking to LIRC and have proven you can control your air conditioner with it, you have lots of interesting options for automating the actual running of the A/C. Go to *http://www.lirc.org/software.html* for a list of LIRC-related and compatible software and applications that let you manipulate your A/C in different ways or integrate it with your multimedia setup. I like using the command line, so my example script will be written in bash. But don't let that stop you from experimenting with everything out there!

Let's say that the room your A/C is in heats up in the daytime while you're at work or school. The air conditioner can be left on to keep the room cool all day, but it would be much more economical to have it turn on right before you come home. So let's say you want it to come on at 4:30 PM. Linux has a daemon called `cron` that can run programs on schedules. You could write a `cron` job to accomplish your goal by entering:

```
$ sudo crontab -e
```

Then add a line like this to your crontab:

```
30 16 * * * /path/to/your/script/air_conditioner.sh
```

This specifies that a script named *air_conditioner.sh* on the path specified will be run at 30 minutes after the 16th hour each day. (If you want to do more with `cron`, try reading `man crontab`!)

Next, create the actual script in your text editor and save it as *air_conditioner.sh*. An example of a bash script to turn on the air conditioner might be as simple as:

```
#!/bin/sh
#
# Sample program to control A/C
#
irsend SEND_ONCE Haier_AC-5620-30 KEY_POWER
```

This assumes that the A/C is always off when you leave in the morning. Since it is an *open-loop* control system (meaning the computer has no way of knowing whether the A/C was on or off to begin with), the beginning state has to be known for it to work right. Therefore, some things you want to do may require cleverness on your part to ensure the starting state is always known.

TAKE IT FURTHER

One option for modifying this project is to make it a *closed-loop* system where the Pi can actually check on the state of the A/C. You could set up a temperature sensor in the room, or an inductive current sensor on the air conditioner's power cord, then have the Pi read the signals from those to determine what the A/C is doing. You could even have the Pi visually check for the state of the LEDs on the front of the A/C, if you are feeling ambitious and have a camera.

Even if you keep it an open-loop system, there are ways to make use of external information to add power to your automation. Here's an experiment you can try right now: say you want the A/C to come on at 4:30 only if it's over 90 degrees outside. You could power it off in the mornings, and have your Pi check the NOAA (National Oceanic and Atmospheric Administration) website for current conditions in your area, like this:

```
wget -q -O current_conditions.txt "http://forecast.weather
.gov/MapClick.php?CityName=Los_Angeles&state=CA&site=LAX&lat
=34.06076&lon=-118.23510"
```

NOTE

Like all websites, NOAA's may change its URL or content, which may break your script. If that happens, modify the URL and/or sed pattern matching in your script to get the information you want.

Unless you live in L.A., you'll want to change the URL to the one for your own city once you have this working. The following commands will strip off the saved website, leaving only the current temperature from the current day:

```
grep -e myforecast-current-lrg current_conditions.txt | sed
's/.*myforecast-current-lrg\">\(.*\)\&deg;F<\/p>.*/\1/g'
```

It's useful to add logging functionality, so you can see if the Pi tried to send a power on command or not. You can do so with the logger command, like this:

```
logger 'Your message here'
```

Putting it all together, I present this sample program for you to edit and play with:

```
#!/bin/sh
#
# Sample program to control A/C
#
CONTROL_TEMP=90
wget -4 -q -O current_conditions.txt "http://forecast
.weather.gov/MapClick.php?CityName=Los_Angeles&state=CA&site
=LAX&lat=34.06076&lon=-118.23510"
CURRENT_TEMP=`grep -e myforecast-current-lrg
current_conditions.txt | sed 's/.*myforecast-current-lrg\">\
(.*\)\&deg;F<\/p>.*/\1/g'`
if [ $CURRENT_TEMP -ge $CONTROL_TEMP ];
then
        irsend SEND_ONCE Haier_AC-5620-30 KEY_POWER
        logger 'Temperature of '$CURRENT_TEMP' is at or above
'$CONTROL_TEMP'. Air conditioner power command sent.'

else
        logger 'Temperature of '$CURRENT_TEMP' is below
'$CONTROL_TEMP'. Air conditioner power signal not sent.'
fi
```

SUMMARY

In this chapter you learned about infrared light and infrared LEDs, and a few—hopefully!—tantalizing bits about communications protocols. You created a system to control an air conditioner remotely from a computer using infrared light. I hope that you'll explore further in books and online because there is a lot to learn about serial communications, circuit design, and LED technology. To learn the nitty-gritty of designing electronics circuits from discrete components, including LEDs, I recommend *The Art of Electronics* by Horowitz and Hill.

STARFIELD LIGHT EFFECT BOX

BY ADAM WOLF

IN THIS PROJECT, YOU'LL MAKE A DECORATIVE DISPLAY BOX WITH A STARFIELD LIGHT EFFECT.

The starfield effect is one of the earliest computer effects demos: stars are rendered on a screen and blurred to make it look like you're flying through them. Stars far away move slowly, and those close to you zoom by. Many people know of this effect from the Windows 3.1 screensaver, but it's been around since at least the late '70s. In this project, we'll create a starfield demo using a 32 × 32 LED matrix, with a SmartMatrix SD Shield to act as intermediate hardware between the microcontroller and the LEDs. This shield microcontroller handles most of the complexity involved in driving all those LEDs.

I've mounted the matrix in a shadow box with a black diffuser. I call it a Bowman Box, after Dave Bowman from *2001: A Space Odyssey*.

GET THE PARTS

I've done my best to make the code and explanation for this project easy to adapt to any LED matrix, but you'll find this project easiest if you use the SmartMatrix SD Shield and a SmartMatrix SD Shield–compatible LED matrix.

Components

- 32 × 32 RGB LED matrix (Adafruit P/N 1484 or SparkFun P/N 12584; must be compatible with the SmartMatrix SD Shield that plugs directly into a Hub75 connector)
- 5 V 4A+ wall power adapter (Adafruit P/N 1466; must have the standard 2.1 mm center-positive connector)
- SmartMatrix SD Shield (available from the manufacturer or Adafruit P/N 1902; v3 is the latest version as of this writing)

NOTE

If you change the project to have more of the LEDs on at once, you'll have to either find a switch rated for more current or plug the power supply into the board directly.

- Teensy 3.1 or 3.2 (Adafruit P/N 2756; this is a small, inexpensive, and really powerful microcontroller board)
- MicroUSB cable
- Inline power switch with 2.1 mm connectors (Adafruit P/N 1125 or SparkFun P/N 11705)
- Shadow box (8 × 8 inches)

- Diffuser (a piece of translucent material that can help diffuse the light, filling in the gap between LEDs; I used a piece of thin black felt, but you could use tissue paper, plastic, or even something smeared on the glass)
- (Optional) Cardboard (to get everything to fit snugly in the shadow box)

Tools

- Soldering iron and solder
- Diagonal cutters
- Wire strippers
- Small screwdriver
- Scissors
- Handsaw, drill, or file
- Black marker

BUILD IT

First, you'll assemble the SmartMatrix SD Shield according to the manufacturer's instructions and connect it to the LED matrix. You'll then need to sandwich the diffuser between the LED display and the shadow box glass. After checking that everything fits nicely into the shadow box, you'll take the box's back off again, cut a small notch for the power cord, upload the program, and reassemble everything.

1. **Assemble the SmartMatrix SD Shield.** Go to the manufacturer's site at *http://docs.pixelmatix.com/* and navigate to the SmartMatrix Shield section. Find the SmartMatrix SD v3 section, and click **Kit Assembly** in the menu on the left. This should take you to your shield's assembly instructions. Read and carefully follow the manufacturer's instructions—this requires a lot of soldering—and then double-check your work.

    With your shield finished, plug it into the LED panel, in the input connector, as shown in Figure 7-1.

NOTE

If you need instructions on soldering, see the appendix.

SMARTMATRIX
SD SHIELD TEENSY LED MATRIX

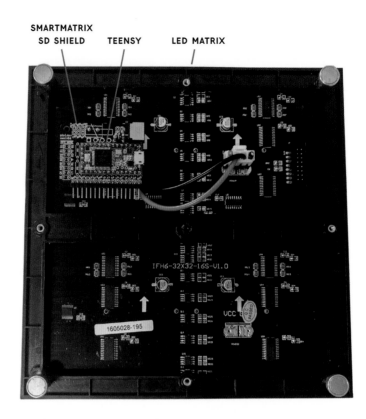

The Smart Matrix SD Shield uses a Teensy 3.1 or 3.2 micro-controller board. We'll program the Teensy in Arduino, which works on Windows, macOS, and Linux.

2. **Set up the Teensy.** Download the latest version of the Arduino IDE from *https://www.arduino.cc/*, install Teensyduino from *https://www.pjrc.com/teensy/teensyduino.html*, and then install the SmartMatrix3 library from the IDE's Library Manager. Connect the Teensy to your computer with a microUSB cable and connect the display to power. Go to **Tools ▸ Board** and choose **Teensy 3.2/3.1**.

3. **Test the Teensy.** Load the *FeatureDemo* example by going to **File ▸ Examples ▸ SmartMatrix3 ▸ FeatureDemo**. Upload the sketch to the Teensy, and you should see the Teensyduino Uploader appear and send the sketch to the Teensy. If you've never used a Teensy before, you may notice that it looks different from the standard Arduino IDE, but the Teensy part all happens automatically so you don't need to do anything extra.

If you have everything working, you should see a demo play out on the Teensy that goes through all the different features of the SmartMatrix3 library. You'll see different colors, moving shapes, and scrolling text. This indicates that the computer, the LED panel, the SmartMatrix SD Shield, and the Teensy are all working. Now let's finish the enclosure.

4. **Cut the diffuser.** Take the back off the shadow box and measure the inner dimensions. Use scissors to cut your diffuser material, as shown in Figure 7-2. You can match the inner dimensions of your shadow box exactly, or you can cut them slightly bigger so you can glue the edges to the sides of the box.

FIGURE 7-2:
Cutting the diffuser

If you're a little off, it's probably okay! The rest of the inside of the box will be black, so any gaps won't be noticeable.

5. **Test-fit the display into the shadow box.** Clean any smudges off the glass of the shadow box and then, with the glass side of the box facing down on a table, place the diffuser in the box and against the glass. Then place the display, LEDs down, on the diffuser. Screw the little magnetic screws into the corners of the LED panel, if you have them. Check to make sure that the back of the shadow box still fits on, as shown in Figure 7-3.

If the fit is too tight, try removing any black velvety material on the back of the shadow box or taking the magnetic screws off the LED panel frame. If the fit is too loose, you can use some cardboard as a shim to keep everything in place.

FIGURE 7-3:

Testing how everything
fits in the shadow box

6. **Cut a notch in the back of the shadow box.** Remove the
   back of the shadow box again. Plug the 2.1 mm inline switch
   power plug into the DC power connector of the SmartMatrix SD
   Shield. The cable should just reach outside of the box. Cut a
   small notch, about 3/4 inch deep by 1/4 inch wide, in the back
   of the shadow box to feed the power cord through, as shown in
   Figure 7-4. I suggest you cut the notch in the middle of an edge
   or at a corner. You can make this notch with many different tools,
   like a handsaw, a drill, a file, or even needle-nose pliers.

FIGURE 7-4:

The back of the shadow
box with the notch cut
out for the power cord

If your shadow box has a velvety backing, you'll need to cut
a slit in that, too, and be sure it aligns with the notch you cut in
the back. Finally, touch up any scuff marks with a black marker.

7. **Assemble it.** Now reattach the back of your shadow box. Everything should be nice and snug. If not, use some cardboard as shims and spacers. Any minor cosmetic issues will be easy to ignore once you get the stars zooming past. On that note, take the back off again so we can access the Teensy for programming!

If you'd like to go directly to the next step, skip ahead to "Code It" on page 140. If you're curious about how the starfield effect works, read on.

If you'd like to go directly to the next step, skip ahead to "Code It" on page 140.

GENERATING THE STARFIELD EFFECT

We need to create the effect of zooming through a bunch of stars. The stars that are close should be brighter and move faster than the stars that are far away. All of the stars should move radially out from the center.

To make the starfield effect in software, we'll create a list of stars and randomly distribute them as points in 3D space with a coordinate system. We'll represent each point as three coordinate values, like (x, y, z), where x represents the horizontal axis, y represents the vertical axis, and z represents depth. You can imagine z as the distance from you—the higher the z value, the farther away the point will appear. Our *eye*, the default point in the 3D space, is at the origin, or $(0, 0, 0)$.

Once we've completed our list of randomly placed stars, we have to overcome the problem of drawing a point in 3D space on a 2D display. Imagine drawing a region of space as a four-sided pyramid with the top and bottom cut off. This is called a *view frustum* and is illustrated here.

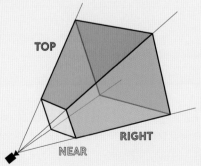

(continued)

The yellow square represents our display, with one LED per unit in x and y. This means that a point at (0, 0, 0) is in the middle of the display, a point at (–8, –8, 0) is in the middle of the lower-left corner, and a point at (8, 8, 0) is in the middle of the top-right corner of the LED panel.

We also need to account for the z distance. A point at (8, 8, 100) shouldn't be drawn in the middle of the top-right corner of our LED panel. It should be closer to the middle of the display than the point (8, 8, 0), because it is farther away from the camera. In other words, the x- and y-coordinates need to move closer to the eye as the z-coordinate gets bigger.

To translate from 3D coordinates to our 2D coordinates, we divide each of the x- and y-coordinates by z. Try thinking about the 2D coordinates as percentages of the total frame size at that particular depth (or z). Dividing the x- and y-coordinates by z gets us a 45-degree angle on our frustum edges, so that for every step in z, we can see an additional step in x and y. We can experiment with different factors, like dividing by double z, or dividing by half of z, and that changes the slope of the edges of our view frustum.

In my Bowman Box, stars at any distance are drawn by a single LED, but the brightness of the LED is relative to the distance: the stars that are the farthest away from us are at the minimum brightness, and the stars that are at the front of the frustum in the very center are at the maximum brightness. To make it simpler in the program, we scale based only on the z-value.

Phew! The programming is much shorter than the explanation! Now that we know how to draw the stars on our display, we need to move! We want to fly through the stars, but instead of moving the eye forward in z, we're going to move all the stars toward the eye in z. It makes the programming easier. Once a star advances past our eye, we certainly can't see it anymore, so we'll recreate it back at the maximum distance away from us.

CODE IT

If you want to skip past all the explanation of the code and how it works, you can simply download the code at *https://nostarch.com/ LEDHandbook/* and upload it to your project. The full code is shown in Listing 7-1.

```
/* StarfieldEffect.ino
   by Adam Wolf

   Uses a SmartMatrix SD Shield, a 32 by 32 RGB LED display,
   and a Teensy to show a starfield effect.
*/

❶ #include <SmartMatrix3.h>
   #define COLOR_DEPTH 24
   const uint8_t kMatrixWidth = 32;
   const uint8_t kMatrixHeight = 32;
   const uint8_t kRefreshDepth = 36;
   const uint8_t kDmaBufferRows = 4;
   const uint8_t kPanelType = SMARTMATRIX_HUB75_32ROW_MOD16SCAN;
   const uint8_t kMatrixOptions = (SMARTMATRIX_OPTIONS_NONE);
   const uint8_t kBackgroundLayerOptions =
    (SM_BACKGROUND_OPTIONS_NONE);
   const uint8_t kScrollingLayerOptions =
    (SM_SCROLLING_OPTIONS_NONE);
   const uint8_t kIndexedLayerOptions = (SM_INDEXED_OPTIONS_NONE);
   SMARTMATRIX_ALLOCATE_BUFFERS(matrix, kMatrixWidth,
    kMatrixHeight, kRefreshDepth, kDmaBufferRows,
    kPanelType, kMatrixOptions);
   SMARTMATRIX_ALLOCATE_BACKGROUND_LAYER(backgroundLayer,
    kMatrixWidth, kMatrixHeight, COLOR_DEPTH,
    kBackgroundLayerOptions);
   SMARTMATRIX_ALLOCATE_SCROLLING_LAYER(scrollingLayer,
    kMatrixWidth, kMatrixHeight, COLOR_DEPTH,
    kScrollingLayerOptions);
   SMARTMATRIX_ALLOCATE_INDEXED_LAYER(indexedLayer, kMatrixWidth,
    kMatrixHeight, COLOR_DEPTH, kIndexedLayerOptions);

   struct Star {
     int16_t x;
     int16_t y;
     float z;
   };

   #define NUM_STARS 32
   struct Star stars[NUM_STARS];

   const int X_RANGE = 400;
   const int Y_RANGE = 400;
   const int HALF_X_RANGE = X_RANGE / 2;
   const int HALF_Y_RANGE = Y_RANGE / 2;
   const int MAX_Z = 16;

   const uint8_t SCREEN_WIDTH = 32;
   const uint8_t SCREEN_HEIGHT = 32;
```

LISTING 7-1:

The starfield effect code

```
    const uint8_t HALF_SCREEN_WIDTH = SCREEN_WIDTH / 2;
    const uint8_t HALF_SCREEN_HEIGHT = SCREEN_HEIGHT / 2;

    const float SPEED = 0.1;

❷  void randomizeStarPosition(struct Star* s) {
      s->x = random(-HALF_X_RANGE, HALF_X_RANGE);
      s->y = random(-HALF_Y_RANGE, HALF_Y_RANGE);
      s->z = MAX_Z; // Put the star in the back.
    }

    void drawStar(struct Star* s) {
      uint8_t display_x = (int) (s->x / s->z);
      uint8_t display_y = (int) (s->y / s->z);

❸    display_x = display_x + HALF_SCREEN_WIDTH;
      display_y = display_y + HALF_SCREEN_HEIGHT;

      // Let's check if our calculated point should be displayed.
      if (display_x >= 0 &&
        display_y >= 0 &&
        display_x < SCREEN_WIDTH &&
        display_y < SCREEN_HEIGHT) {
        uint8_t brightness = map(s->z, 0, MAX_Z, 255, 0);
        drawPoint(display_x, display_y, brightness); // It fits,
                                                     // so draw it!
      } else {
        randomizeStarPosition(s);  // It does not fit,
                                   // so make a new star.
      }
    }

    void advanceStar(struct Star* s) {
      s->z -= SPEED;

      if (s->z <= 0) {
        randomizeStarPosition(s);  // If the star has passed us,
                                   // make a new one.
      }
    }

    void setup() {
❹    randomSeed(analogRead(A1));

      for (int i = 0; i < NUM_STARS; i++) {
        Star* current_star = &stars[i];
        randomizeStarPosition(current_star);
❺      current_star->z = random(0, MAX_Z);
      }
```

```
    setupDisplay();
  }

  void loop() {
    clearDisplay();

    for (int i = 0; i < NUM_STARS; i++) {
      drawStar(&stars[i]);
    }

    paintDisplay();

    for (int i = 0; i < NUM_STARS; i++) {
      advanceStar(&stars[i]);
    }
  }

  uint16_t distanceToStar(struct Star* s) {       // distance
    return sqrt(sq(s->x) + sq(s->y) + sq(s->z)); // from (0, 0, 0)
  }

  void clearDisplay() {
    backgroundLayer.fillScreen({0x00, 0x00, 0x00});
  }

  void drawPoint(uint8_t x, uint8_t y, uint8_t brightness) {
    // Note this doesn't actually show it on the display.
    backgroundLayer.drawPixel(x, y,
      {brightness, brightness, brightness});
  }

  void paintDisplay() {
    // Send the image we've built up to the screen.
    backgroundLayer.swapBuffers(false);
  }

❻ void setupDisplay() {
    matrix.addLayer(&backgroundLayer);
    matrix.addLayer(&scrollingLayer);
    matrix.addLayer(&indexedLayer);
    matrix.begin();

    backgroundLayer.enableColorCorrection(true);
  }
```

As you're reading this sketch, there are a few things to notice. Arduino sketches run the `setup()` function first, and then repeatedly run the `loop()` function.

The blocks of code at ❶ and ❻ are taken directly from the SmartMatrix examples that come with the IDE, and are used to set up the SmartMatrix library.

We've used a *struct*, which is a grouping of variables to store some of the information in our sketch. Structs can make it easier to see the intent of your code by grouping related information together. We make a bunch of star structs at the beginning of the program. When a star has advanced behind our eye or outside our viewing range, we randomize that star's coordinates and then set it far away from us so we can encounter it again ❷. This lets us reuse the structs rather than needing an infinite quantity.

We want the origin of the stars to be in the middle of the display, but the SmartMatrix library sets the origin at the bottom-left corner, so next we adjust for that when we display the stars ❸.

We also need a random-number generator to place the stars in random positions. The Arduino doesn't have a real random-number generator, so we initialize the system by reading a disconnected pin at ❹ to get different pseudorandom numbers each time. Reading a disconnected pin is not a very good "random seed," so if you run it multiple times you may see it start up the same way. There are ways to make this more random, but we've kept it simple for clarity.

Because normally new stars show up only at the back, when we first initialize the position of the stars, we scatter them throughout the z-axis ❺.

Make sure you've connected the power supply to the power switch, and then turn the switch on. Upload the code to your Teensy, make sure everything uploaded properly, and reattach the back on your shadow box.

SUMMARY

Congratulations! You should now have a beautiful LED matrix mounted in a nice box. You can improve and expand this project in many ways. But remember that in our project, the majority of the LEDs are off at any time. If you change the project and turn a lot more LEDs on at the same time, you'll need to be careful with power. The inline switch is not rated to carry enough current to light the whole screen on full brightness, but you can remove it and plug the power supply into the board directly.

As a further upgrade, I've considered adding an accelerometer so I can steer by moving the frame, changing the angle and the speed as I fly through the stars. I might even add colored stars—not bright rainbow ones, but slightly red and slightly blue stars, to represent planets and nebulas. The sky's the limit! Try a few things out and see which effect suits you best.

OPTICAL SCREWDRIVER

BY JOHN BAICHTAL

THIS PROJECT IS A LIGHT-BASED BEATBOX IN WAND FORM, USED WITH A LIGHT-SENSOR-EQUIPPED SYNTH TO CREATE COOL AUDIO EFFECTS.

In this project, you'll make an optical screwdriver that flashes an LED at a light sensor, which then tells the synth what beats to play depending on the input it receives. You can modify the rate and length of beats, creating complicated synthesized music.

I'll show you how to make three different variants of the screwdriver, ranging from simple to advanced. You don't have to make all three; just choose the level of awesomeness you're prepared to handle.

The first version is a super-basic Arduino project easy enough for anyone to build. You'll also need to assemble a test rig—an elementary synth consisting of an Arduino, a speaker, and a light sensor—to make sure it works. After that I'll show you how to level up your screwdriver with a couple of improvements: a simplified microcontroller and a wand-shaped printed circuit board (PCB).

BUILD THE BASIC SCREWDRIVER

Let's begin with the easiest possible configuration for the optical screwdriver: a plain-jane Arduino with an LED and three potentiometers, as shown in Figure 8-1. While not particularly wand-like, it gives you an unsexy equivalent that works identically—it makes optical LED flashes that interact with a sensor-equipped synthesizer. The three pots will be used to control the speed, duration, and pauses of the flashing LED.

FIGURE 8-1:

A basic version of the screwdriver

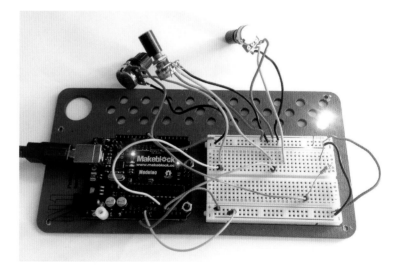

Get the Parts

You'll need just a few parts to complete the first stage of this project. I'm also including the parts for the test rig you'll need to build in order to hear anything. The test rig is simply a light sensor, Arduino, and speaker.

- Two Arduino Uno boards or similar (Adafruit P/N 50)
- Half-sized breadboard (Adafruit P/N 64)
- Jumper assortment (Adafruit P/N 153)
- Wall wart or 9 V battery clip (Adafruit P/N 63 or Adafruit P/N 80)
- Super-bright LED (for example, SparkFun P/N 531)
- Three 10 kΩ potentiometers (I used SparkFun P/N 9939, but P/N 9288 also works; I used slim knobs from Adafruit, P/N 2057)
- Light sensor (Adafruit P/N 161)
- 8 Ω speaker (for example, Adafruit P/N 1313; you'll need to solder on your own wires)
- 220 Ω and 10 kΩ resistors (SparkFun P/N 10969 is a good multipack)

NOTE

See "Getting Started with the Arduino and the Arduino IDE" on page 15 for setup instructions.

Assemble the Circuit

The bare-bones screwdriver probably won't challenge you, but never fear—this project levels up quick! Let's begin:

1. **Plug in the LED.** Connect the long lead of the LED to pin 9 with a 220 Ω resistor in between, then connect the short lead to GND right next to it. Figure 8-2 shows how it should look.

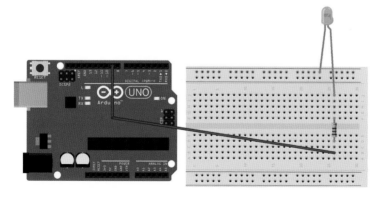

FIGURE 8-2:
Adding the LED and resistor to the breadboard

2. **Wire up the potentiometers.** Add the three pots. For each pot, connect the leftmost lead to GND and the rightmost to 5V.

The center lead is the data connection, as shown in Figure 8-3. It doesn't matter which lead is GND and which is 5V, as long as the center lead goes to data. Connect one of each wire to pins A0, A1, and A2, shown as purple wires in Figure 8-3.

FIGURE 8-3:

Connecting the pots to the appropriate analog pins

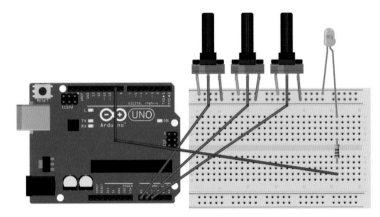

3. **Add power.** Connect the Arduino to power by plugging it into either a wall wart or a 9 V battery. You'll also need to power the prototyping board, so connect the 5V and GND pins on the Arduino to the power and ground buses on both sides of the board. These new wires are shown in red (power) and black (ground) in Figure 8-4.

FIGURE 8-4:

Connecting the breadboard to the Arduino's 5V and GND pins

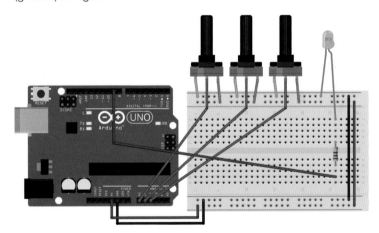

Code It

The sketch is essentially the Blink example sketch, to which I've added potentiometers to modify the beat. You can download it along with the other project files for this book at *https://nostarch.com/ LEDHandbook/*.

Let's go through the code part by part:

```
void setup() {

  pinMode(9, OUTPUT); //change to 0 with the Tiny85
  pinMode(A0, INPUT);
  pinMode(A1, INPUT);
  pinMode(A2, INPUT);
}
```

NOTE

Remember to change all instances of pin 9 to pin 0 if you employ the ATtiny85 microcontroller—more on this later!

The void setup() function runs once when the Arduino is powered on or reset. In this case the setup is telling the Arduino which pins will be for input and which for output.

Next is the loop.

```
void loop() {
  int speedKnob = analogRead(A0);
  int durationKnob = analogRead(A1);
  int skipKnob = analogRead(A2);

  int currentSpeed = map(speedKnob, 0, 1027, 10, 900);
  int currentDuration = map(durationKnob, 0, 1027, 100, 800);
  int currentSkip = map(skipKnob, 0, 1027, 1, 150);
```

The first part of void loop() declares a series of variables corresponding with the three pots—speed, duration, and skip. For the sake of writing code, I've labeled time between beats as *speed*, the length of each beat as *duration*, and the brightness level of the LED in the "off" state as *skip*. This gives us the option of "bright to dim" rather than "on or off," for a smoother flow.

Finally, the code maps out the readings from the three pots and changes their values to something we can use—data comes in from the pots at a range of 0–1023, so the code translates that value into seconds. Feel free to change these values to make the screwdriver work better for you!

```
  digitalWrite(9, HIGH); //the LED turns on
  delay(currentDuration);
  digitalWrite(9, LOW); //the LED turns off for a moment
  delay(10);
  digitalWrite(9, currentSkip); //the LED dims
  delay(currentDuration * 0.5);
  digitalWrite(9, LOW); //the LED turns off
  delay(currentSpeed);
}
```

The final part of the code governs the actual work of the sketch: turning on and off an LED. The LED turns on for a number of microseconds equal to the value of `currentDuration`, or between 100 and 800. When the synth's light sensor detects those beats, it responds with sound. Holding the screwdriver closer makes louder beats, while pulling it back yields more subtle sounds.

Build the Test Rig

Unless you have a light-detecting synth handy, you'll have to build one. This rig emits a beat that is modified by data sent via the light sensor, so it can interact with the screwdriver. You'll need the extra Arduino from the parts list, as well as the speaker. You can see the ultra-simple test rig I built in Figure 8-5.

1. **Wire up the speaker.** Plug the speaker's positive lead into pin 9 of the second Arduino, and the negative lead into GND. Figure 8-6 shows how it connects.

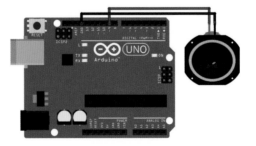

2. **Attach the light sensor and resistor.** Connect one lead of the light sensor to pin A0 and the other to 5V. Slide a 10 kΩ resistor into the same A0 pin, or wrap the light sensor lead and resistor

together. Connect the other lead of the resistor to GND, as
shown in Figure 8-7.

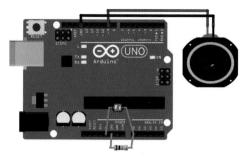

FIGURE 8-7:
Attaching the light sensor
and resistor

3. **Code it.** The test rig's code is a variant of the `tonePitchFollower`
    example sketch, which changes a tone based on the level
    detected by the light sensor. I've added a bit more pizzazz by
    tying the beat length and pause length to the same light sensor,
    allowing the optical screwdriver to simultaneously modify the
    beat's tone, its length, and the time between beats via the light
    sensor.

```
void setup() {
}

void loop() {

   int sensorReading = analogRead(A0);
   int thisPitch = map(sensorReading, 400, 1000, 120,
1500);
   int reverseReading = map(sensorReading, 400, 1000,
1023, 1);

   tone(9, thisPitch, reverseReading);
   delay(10);

}
```

4. **Test it.** The test rig creates a looping series of tones, 100 per
    second, with a frequency equal to the light sensor reading, and
    determines the duration of tone by remapping the sensor read-
    ing. Even without the screwdriver, you can play with the test
    rig by putting your finger over the light sensor and changing its
    reading.

When you add the screwdriver's weirdly flashing light into the mix, even more complicated patterns emerge as the fluctuations in light create complex sonic structures.

BUILD THE WAND VERSION WITH AN ATTINY85

Next I'll show you how to swap in an ATtiny85, a smaller microcontroller than the ATmega328p featured in the Arduino. This is a simpler chip, without the extra bells and whistles of the Arduino board, but it's also more compact, which allows you to move the screwdriver circuit off the prototyping board and onto a more portable wand.

Get the Parts

Grab the following parts to build a soldered version of the basic screwdriver using the ATtiny85. If you want to, reuse the parts from the earlier prototype build.

Components

- Arduino Uno (reuse the one from "Build the Test Rig" on page 152)

- ATtiny85 microcontroller (for example, SparkFun P/N 9378)

- Super-bright LED (for example, SparkFun P/N 531)

- Two 220 Ω resistors

- 10 µF capacitor (SparkFun P/N 523)

- Perma-Proto board (Adafruit P/N 571; these boards require you to solder the components)

- 8-pin IC socket (Adafruit P/N 2202)

- Three potentiometers (I suggest SparkFun P/N 9288 because they can be soldered in place)

- 5 V power supply (for example, SparkFun P/N 12889)

- Barrel jack for the power supply (Adafruit P/N 373 or SparkFun P/N 10811)

NOTE

You shouldn't use a 9 V battery because the ATtiny85 lacks the power management features of the Uno, which allows input voltage to be converted to 5 V. If you use a 9 V battery, your chip will overload.

NOTE

See the appendix for soldering instructions.

Tools

- Soldering iron

- Solder

Let's take a look at the technical details of the ATtiny85. It has eight pins, consisting of power, ground, and a variety of analog and digital data pins. Figure 8-8 breaks down the pin layout.

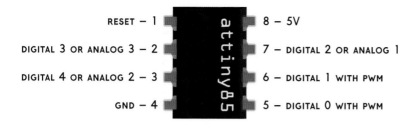

RESET — 1

DIGITAL 3 OR ANALOG 3 — 2

DIGITAL 4 OR ANALOG 2 — 3

GND — 4

attiny85

8 — 5V

7 — DIGITAL 2 OR ANALOG 1

6 — DIGITAL 1 WITH PWM

5 — DIGITAL 0 WITH PWM

FIGURE 8-8:
The ATtiny85's eight pins control a surprising amount of computing power.

Flashing the ATtiny85

One downside to using the ATtiny85 is that we lose some of that ease of programming we have with the Arduino. The Arduino has a lovely USB-based bootloader, but its smaller cousin does not, so you'll have to use a separate Arduino to program the ATtiny85. Since the pins of the chip are needed to program it, we can't assemble the project before programming it. With this version of the optical screwdriver, you won't be able to reflash the chip once it's installed on the circuit board.

Here's how to do so:

1. **Wire up the ATtiny85.** Use the Arduino from "Build the Test Rig" on page 152 as shown in Figure 8-9. The pinout shown in Figure 8-8 becomes important in this step as you wire it up. Connect pin 1 to digital pin 10, shown as a blue wire in Figure 8-9. Connect pin 4 to GND via a black wire. Plug a red wire from pin 5 into a power bus. Plug pins 6, 7, and 8 into 13, 12, and 11 on the Arduino, shown as yellow, white, and green wires, respectively.

2. **Connect the LED and capacitor.** You'll also need to connect an LED, with a 220 Ω resistor protecting it, to pin 8 on the ATtiny85; this LED will flash to show that it's working. Next, you'll need to connect the short pin of a 10 µF capacitor to the Arduino's Reset pin (shown as a burgundy wire) and the long pin to GND.

3. **Connect the power and ground.** Finally, connect the breadboard's power and GND buses to their respective Arduino pins.

4. **Install the ArduinoISP sketch on the Uno.** Find this in the Tools menu of the Arduino program, then set the Uno's programmer to "ArduinoISP" by choosing **ArduinoISP** from the list of programmers. Go to **Tools ▸ Board** and choose the correct ATtiny model. Then upload the sketch!

You can also buy products that allow you to program your ATtiny85 without needing a second Arduino—for example, the Tiny

AVR Programmer from SparkFun (P/N 11801) and the ISP Shield from Evil Mad Scientist (*http://emsl.us/253*).

This is a necessarily stripped-down description of the process. For a more thorough guide to programming an ATtiny, go to MIT's High-Low Tech page at *http://highlowtech.org/*, click **Tutorials**, then click the link for **Programming an ATtiny w/ Arduino 1.6 (or 1.0)**. David Caminati also has a good tutorial at *http://fritzing.org/projects/programmer-for-attiny85-with-arduino-uno-as-interf/*.

FIGURE 8-9:

Programming the ATtiny85 with another Arduino

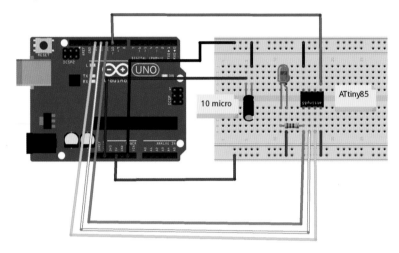

Build the ATtiny85 Screwdriver

The next challenge involves turning that loose jumble of wires into something a little more solid. You'll use a credit card–sized prototyping board from Adafruit and design it so that, besides a power supply wire, you can hold the entire screwdriver in one hand.

1. **Solder in the pots.** Solder the three potentiometers to the prototyping board, as shown in Figure 8-10. Connect the outer two leads of each pot to power and ground.

FIGURE 8-10:

Attaching the potentiometers

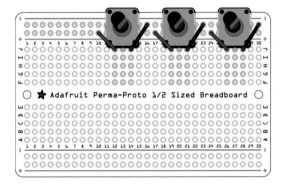

2. **Attach the IC socket.** Next, solder in the IC socket as shown in Figure 8-11. The top of the ATtiny85 has a notch that should be pointed toward the star, as shown in the figure. Connect pins 2, 3, and 7 to the center leads of each pot, from right to left, respectively, and connect pin 8 to power and pin 4 to ground (refer back to Figure 8-8 for the ATtiny85 pinout). If you get these in the wrong order, you might get unexpected results, so make sure to take note of which pot is connected to which pin.

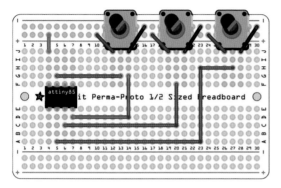

FIGURE 8-11:

Soldering in the IC socket

3. **Wire up the LED.** Connect the long lead of the LED to pin 5 of the ATtiny85 via a 220 Ω resistor, and connect the short lead of the LED to ground. Figure 8-12 shows how it should look.

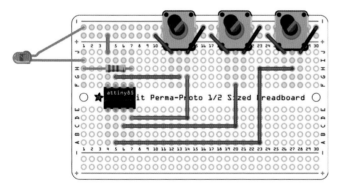

FIGURE 8-12:

Adding the LED and resistor

4. **Add power.** Solder in the barrel jack (shown in Figure 8-13), connecting the center lead to power and either of the two side leads to ground.

As long as you've already flashed the ATtiny85 with the sketch, you should be able to plug it in right away.

FIGURE 8-13:

Soldering in the
power jack

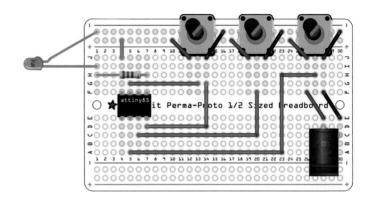

That's your circuit all soldered!

Code It

The code from the initial build works with this version too! You do,
however, have to make one critical change. There is no pin 9 on an
ATtiny85, so you have to change that to a 0. Remember this line from
the original code?

```
pinMode(9, OUTPUT); //change to 0 with the Tiny85
```

You need to change that 9 and the following 9s to 0s.

```
digitalWrite(9, HIGH); //the LED turns on
delay(currentDuration);
digitalWrite(9, LOW); //the LED turns off for a moment
delay(10);
digitalWrite(9, currentSkip); //the LED dims
delay(currentDuration * 0.5);
digitalWrite(9, LOW); //the LED turns off
delay(currentSpeed);
```

When you've made the changes, upload the sketch using the
programming rig mentioned earlier in the chapter—that is, by wiring the ATtiny85 to the Arduino and then loading the code to the
Arduino. Once the ATtiny85 is programmed, it can be placed in the
socket and you'll be ready to go.

MAKE A PCB WAND

Without the need for a full-sized Arduino, you can make the optical
screwdriver much smaller and, well, sexier. In fact, you can make it
small enough to fit on a wand-shaped printed circuit board (PCB),

shown in Figure 8-14. The previous version of the screwdriver was for folks who didn't want to buy or mill their own circuit boards. If you're up to that challenge, however, this variant of the project is for you!

ABOUT FRITZING

Fritzing is a simple circuit design program that allows you to build circuits either through traditional schematic symbols, bread-boarded components, or PCB design. To create a design, you simply drag components out of a palette and draw traces between them. I use the program all the time, and you'll probably find it useful as well. Download it at *http://www.fritzing.org/*.

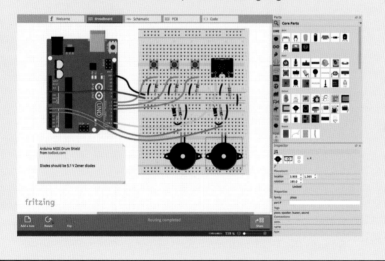

Get the Parts

Most of these parts are the same as the components we used in the previous version of the project. You'll need the following to complete this version.

Components

- Super-bright LED (for example, SparkFun P/N 531)
- ATtiny85 (SparkFun P/N 9378)

- 8-pin IC socket (Adafruit P/N 2202)
- Three 10 kΩ potentiometers (SparkFun P/N 9939 or P/N 9288)
- Mini power switch (SparkFun P/N 102)
- Battery clips (SparkFun P/N 7949)
- Surface-mount LED (SparkFun P/N 12619)
- Two surface-mount resistors (Jameco P/N 2008882)
- Two AA batteries

Tools

NOTE

See the appendix for soldering instructions.

- Soldering iron
- Solder

Build It

Follow these steps to build your optical screwdriver wand, shown in Figure 8-15.

FIGURE 8-15:

The completed optical screwdriver in use

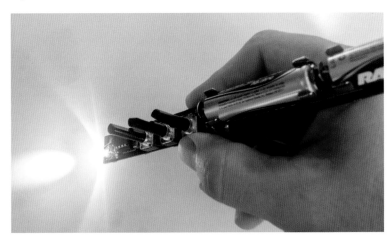

1. **Design the PCB outline.** I used Inkscape (*https://inkscape .org/*) to draw a vector shape corresponding with the shape I wanted the PCB to be. Figure 8-16 shows what I came up with: a pencil shape about 8 inches long and 3/4 inches wide. The Inkscape file is available with all the other resources for this book (*https://nostarch.com/LEDHandbook/*). You can make your circuit board any shape you want. In the Inkscape file, the PCB can be any color, but this won't affect how it looks in the end; what the program looks for are the vectors. Fritzing offers a tutorial on how to make a shape that will work with the program at *http:// fritzing.org/pcb-custom-shape/*.

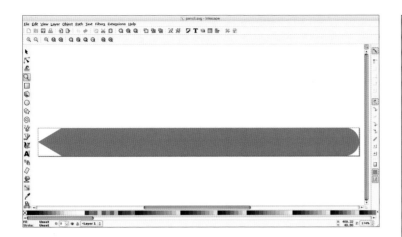

FIGURE 8-16:
I designed a pencil shape in Inkscape for the wand.

2. **Make the circuit in Fritzing.** Add the circuit to the board shape, as shown in Figure 8-17. Don't worry—I'll describe how to do each substep!

FIGURE 8-17:
The optical screwdriver added to the circuit board design

   a. **Add the PCB outline.** To add the circuit board outline you made in Inkscape, click once on the default PCB, then click the **Load Image File** button in the Inspector to replace it with your own. Figure 8-18 shows the PCB outline, ready to be loaded with components. Remember that the blue color doesn't mean anything; it's just a placeholder. The real decision on color is made when you pay for the PCB to be milled.

FIGURE 8-18:
The PCB file has been loaded.

   b. **Add batteries.** Drag battery clips out of the parts palette and add them to the circuit board. Connect the two batteries in series as you see in Figure 8-19. The black lines in the figure represent screen printing that will appear on the PCB, while the circles represent holes drilled in the PCB, with green representing populated terminals and red signifying unpopulated terminals.

FIGURE 8-19:
Adding the battery clips

c. **Add the switch.** The switch has three mounting holes (see Figure 8-20); you should connect the battery to the middle one.

d. **Add the surface-mount LED and resistor.** You can see from Figure 8-21 that the LED and resistor don't have mounting holes, but rather mounting pads. That means these components are surface-mount rather than through-hole. Connect the top lead of the switch to the first pad of the LED, connect the LED and resistor in series, and then ground the resistor to the negative terminal of the battery assembly. This resistor serves as a power indicator, so no matter what happens with the screwdriver you'll know if the batteries are working!

e. **Add the three pots.** Drag the three pots from the palette and add them as shown in Figure 8-22. The left-hand pin connects to power via the switch, and the right-hand pin connects to GND. The center (data) pins remain unpopulated for now.

f. **Add the ATtiny85.** Add the ATtiny85 socket by connecting the pins as you did in the previous project: pins 2, 6, and 7 connect to the pots, while pin 4 connects to ground and pin 8 to power. Figure 8-23 shows how it should be wired up.

g. **Add the resistor and super-bright LED.** The final piece involves the LED and the resistor that protects it. Add the components in series, with the resistor attached to pin 5 of the ATtiny85 and the negative lead of the LED connecting to ground. Figure 8-24 shows the completed PCB.

FIGURE 8-24:
The PCB is complete.

3. **Output the Gerbers.** Export the Gerbers, the design files used by the fabrication staff to mill the boards. Go to **File ▶ Export ▶ For Production** and select **Extended Gerbers**. You will be asked to save the files to a folder.

When the Gerber files are output, check them on a Gerber checker like MCN (Figure 8-25), available from *http://mcn-audio.com/*. This tool lets you inspect the traces visible on the PCB to ensure they are wired as expected. I've had many PCBs ruined by foolishly trusting that the software got it right, so I definitely recommend it.

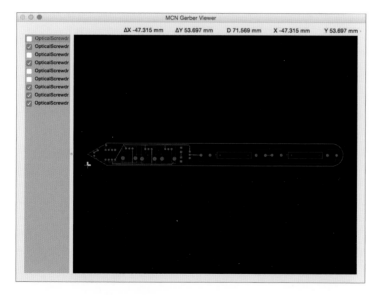

FIGURE 8-25:
Check your Gerbers with MCN or a similar application.

4. **Get the PCB milled.** Chose one of a thousand PCB milling services. I use OSHPark (*http://www.oshpark.com/*) because they're easy and inexpensive. Figure 8-26 shows my milled circuit boards, back from the service.

FIGURE 8-26:

The circuit boards are done. Now to add components!

5. **Solder in the components.** Solder in the components as you would expect. Of particular importance is the ATtiny85, which is susceptible to damage. Rather than soldering the IC directly into the circuit board, solder in the socket and place the IC in the socket. This also allows you to refresh the ATtiny85 at a future date.

6. **Code it.** Upload the optical screwdriver sketch to the ATtiny85 as described earlier in the chapter. It should work the same! Your completed project should look like Figure 8-27.

SUMMARY

Considered by many to be the simplest form of electronics project, a blinking LED nevertheless can offer some cool challenges and opportunities. This project shows how combining digital and analog makes for an intriguing tool that you can make yourself.

WEARABLE TIMING BRACER

BY MIKE HORD

IN THIS PROJECT, YOU'LL MAKE A GLOWING WRIST TIMER THAT'S PERFECT FOR A LARP COSTUME.

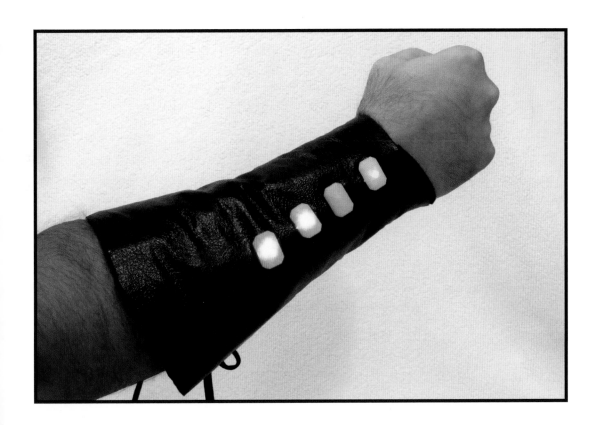

I have a lot of friends who participate in *live action roleplaying (LARP)* games. As a concept it's similar to tabletop roleplaying, except there are no turns, there's no dice, and everything happening around you has a component of reality to it.

One of the elements of the game is timed effects; for instance, a spell may put you to sleep for 10 minutes, pin you in place for 1 minute, or make you laugh uncontrollably for 5 minutes. Since the game takes place in a fantasy wonderland, using a wristwatch to time an effect is an anachronism that detracts from the sense of the game. Players need a way to keep time that's aesthetically unobtrusive. To address this, in this project we'll build a bracer (or wristguard) with mystical glowing gems that also tells you how much longer an effect will last. As a bonus, we'll make it connectable to your smartphone, so you can set the timer duration without reprogramming the project!

The bracer has four gems on it, each of which, when pressed, starts a timer. By default the timers last for one, five, or ten minutes. The gems pulse-glow for the duration of the effect, turning off when the timer runs out.

GET THE PARTS

This project includes both sewing and electronics. I've separated the supplies into two lists for convenience. You can find the craft supplies at any reasonably well-stocked sewing supply or fabric store.

Components

- LilyPad Simblee board (SparkFun P/N 13633)
- 4 LilyPad pixel boards (SparkFun P/N 13264)
- 4 LilyPad button boards (SparkFun P/N 08776)
- 1,000 mAh lithium polymer ion battery pack (SparkFun P/N 13813)
- LilyPad FTDI basic board (SparkFun P/N 10275)
- Mini-B USB cable (SparkFun P/N 11301)
- Conductive thread (SparkFun P/N 13814)

Crafting Supplies

- Grommet kit
- Ribbon, lace, or other attractive string-like material; I'm using parachute cord

- Sundry sewing supplies, like thread and a sewing needle
- Flexible fabric glue
- Clear or translucent gemstones or beads, about 1/2 to 1 inch in size
- Used gift card or expired credit card
- 5 to 10 precut sheets of felt (9 × 12 inches)
- 1/3 yard of faux leather

Tools

- Hammer (for setting grommets)
- Scissors

BUILD IT

Since the measurements for the crafting portion will vary according to preference and the wearer's size, much of this project requires a "try it and see" approach.

I'm also assuming that you're comfortable enough with sewing that I don't need to explain every step of the sewing portion. If you've never done an e-textiles project before, I suggest that you first check out these excellent tutorials to help with the fundamentals:

- ***https://www.sparkfun.com/tutorials/313*** This very useful video tutorial explains basic e-textile stitching practice.
- ***https://www.sparkfun.com/tutorials/306*** We'll use the technique in this tutorial to make the buttons that activate our timers.
- ***https://learn.sparkfun.com/tutorials/dungeons-and-dragons-dice-gauntlet*** This project makes a similar style of costume garment, but uses it to conceal a dice-rolling mechanism instead of a timer.

The final bracer will have three layers: the top layer is the leather-like material, the middle layer contains the piece of material with the circuitry, and the bottom layer is another layer of felt or fabric. Let's get started.

1. **Cut out the fabric pieces.** Figure 9-1 shows the approximate shape you need. The exact measurements depend on how big you want your bracer to be, although you should try to make it

as big as possible to accommodate the electronics. Try a few different shapes and dimensions using paper or scrap fabric until you find one that feels right.

Once you find the shape and size you like, cut out two pieces of that shape from felt and one piece from faux leather. Make the leather piece slightly bigger all around, so it conceals the felt pieces beneath it.

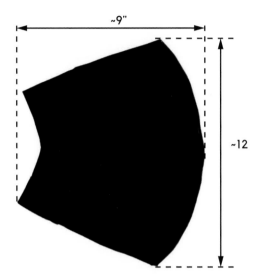

2. **Lay out some electronics.** Now you'll experiment with how to place your electronics on the felt bracer shape. You need to find a position that feels comfortable for you.

First, you need to orient the LilyPad Simblee board with the six-pin header pointing toward an edge, so you can access these pins for programming and charging. Second, the pixel boards (I'll refer to them as LEDs for short) should form a straight line down the center of the bracer, for aesthetic reasons. Figure 9-2 shows how I set up mine.

At this stage, you're laying out only the LEDs, the battery, and the Simblee board. You'll add the buttons later.

FIGURE 9-2:
Line up the electronics in the middle of the form.

3. **Sew in the conducting power threads for the LEDs.** Using conductive thread, first you'll sew in place the positive (+) and negative (−) power threads for the LEDs. You'll need one power thread along one side of the electronics connected to all the positive pads, and one along the other side connected to all the negative pads. See Figure 9-3 for an idea of the threading.

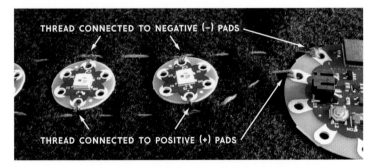

THREAD CONNECTED TO NEGATIVE (−) PADS

THREAD CONNECTED TO POSITIVE (+) PADS

FIGURE 9-3:
The positive and negative power threads for the LEDs

Pay attention to the orientation of the LEDs! Make sure you sew the negative pad on the Simblee to the negative pads on the LEDs. When you're certain you have everything laid out correctly, sew the pads in place with the conductive thread, making sure the stitches are long and have a fair amount of space between them like in Figure 9-3. In the next step you'll have to pass another thread over the negative thread, and it's easiest to do that by passing it between two stitches. The piece of felt that you sew the electronics onto is going to be the middle layer, just under the faux leather piece.

4. **Sew the data thread from the Simblee board to the LEDs.** Now you'll need to link the LEDs to the Simblee board, so the Simblee can control them. Figure 9-4 shows how to sew the conductive thread.

FIGURE 9-4:

Connecting the Simblee board to the LEDs

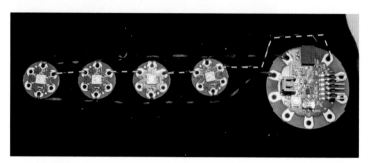

In Figure 9-4 I've used dashed yellow lines to more clearly show the path of the data thread from the Simblee board pad, labeled pad 3 on the board, to the first LED, and the threads that connect the LEDs consecutively.

Note that the data thread crosses over the negative power thread, as I mentioned in the previous step. You must make sure that the two threads don't touch, or your project will short out. To accomplish this, pass the data thread over the negative power thread in the space between two stitches.

5. **Cut holes in the felt under each LED board.** Each LED needs one button under it so that when you press down on the LED it actuates the switch. To make the best connection, cut a small hole in the felt under the center of each LED so the LED board touches the button directly, without any felt between. But take care not to cut too large a hole; you don't want to accidentally cut through one of the threads you've already placed, or weaken the felt around one so that it tears through!

6. **Place, sew down, and thread the button boards.** Place one button board beneath each LED so the button is facing the back of the LED board. There should be exactly one button board under each LED, as shown in Figure 9-5.

Again, I've highlighted the thread paths with yellow. The long thread across the top is the negative power thread from the other side of the felt, which you laid down earlier.

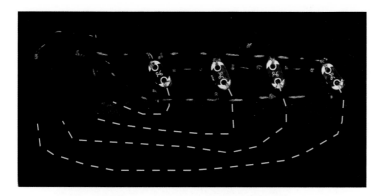

FIGURE 9-5:

Threading the button
boards to the bracer

Stitch conductive thread from one side of each button board
to the negative power thread that you laid down earlier. Unlike
LED boards, the button boards don't have polarity, so you can
thread either end to the negative power thread.

Next, run a thread from the other side of each button to the
Simblee board, as shown in Figure 9-5. Stitch the closest button
to pad 15, the next to pad 12, the next to 11, and the last to 9.
Be extremely careful about where these threads cross the positive power thread from earlier. You need to make *very* certain
that the threads don't touch; otherwise that button won't work,
and it'll be a huge inconvenience to fix.

7. **Glue the backing material to the button boards.** Now cut
   up the old gift card, or some similar stiff material, into squares
   approximately the size of the LED boards' diameter. Glue one of
   these to the back of each button board, as shown in Figure 9-6.

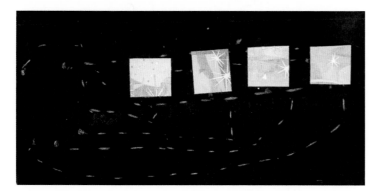

FIGURE 9-6:

The squares of card
material glued to the back
of each button board

I used a fabric glue specially designed to flex. These squares
provide a solid backing for the buttons to press against so they
will actuate rather than sinking into the layers below them. This
step is important and should not be skipped.

8. **Stitch together the layers and place the grommets.** The piece of material you've been working with forms the middle layer; the top layer is the leather-like material, and the bottom layer is another layer of felt or fabric. Stitch the three layers together along one edge, as shown in Figure 9-7. Be sure to use regular, nonconductive thread.

FIGURE 9-7:

Sewing the layers together and placing the grommets

Once you've stitched the edge of the layers together, you can place the grommets on that edge. Follow the instructions on the grommet kit package for these. Then repeat the sewing and grommeting on the other side.

9. **Cut holes over the LEDs.** Expose the LEDs by cutting a small hole in the leather over each one. If you'd like, you can also glue down a translucent gemstone or bead over each hole for a flashier appearance. I used 3D-printed gemstones with a mushroom backing. If you want to 3D-print these gemstones yourself, you can find the files in the book's resources at *https:// nostarch.com/LEDHandbook/*. Thanks to Youmagine user jensa (*https://www.youmagine.com/jensa/designs*) for providing the original "emerald gemstone" design. Figure 9-8 shows the completed project.

FIGURE 9-8:
The complete bracer with gemstones in place over the LEDs

When the bracer is ready to wear, thread your ribbon or cord through the grommets to lace the bracer onto your wrist.

Pressing on an LED, or a translucent decoration if you've added them, should produce a nice "click" as the button actuates. If it doesn't, add more layers of card to the back of the button board until the button is sufficiently sturdy.

CODE IT

Let's take a look at the code. The full code is quite long—more than 500 lines—so I've included only the most interesting parts of it here. Download the code in the book's resources at *https://nostarch.com/ LEDHandbook/*, and read the comments in the following code listing to see how it works in general.

Support for the Simblee is not built into the Arduino IDE, so you'll need to add it yourself. Follow the instructions for doing so in the Setting Up Arduino tutorial at *https://learn.sparkfun.com/tutorials/ simblee-concepts*. That page will walk you through all the pertinent steps for programming the Simblee. This is where you'll need the FTDI basic board and the USB cable.

The code is a lot simpler than its length would suggest. It sets up four independent timers, which can be triggered or reset by pressing the buttons hiding under the LEDs, then creates a pulsing effect on the LED corresponding to a running timer. The lion's share of the code is actually the same code, replicated four times, once for each timer. At the end is all the necessary UI setup code.

```
#include <SimbleeForMobile.h>

#define SFM SimbleeForMobile // A macro to shorten our later
                             //   library calls to a
                             //   reasonable length.

// Aliases for the pin assignments for each button.
#define BUTTON1 15
#define BUTTON2 12
#define BUTTON3 11
#define BUTTON4 9

// Linearize the output of the LED in 16 steps. Because of the
//   nonlinear response of the human eye, for the LED to look
//   like it is fading in a linear fashion, nonlinear steps
//   of brightness are needed.
const int LEDSteps[16] = {255, 191, 154, 128, 107, 90, 76, 64,
                           53,  43,  34,  26,  19, 12,  6,  0};

// Timer variable definitions.
int timer1 = 60;    // timer1 defaults to 60 seconds

int t1Default = 60;

long t1last = 0;

int t1LEDIndex = 0;

int t1LEDDir = 1;

int t1LEDVal = 255;

long LEDUpdateLast = 0;

// Timer active definitions.
bool t1active = false;

// Flash storage definitions and variables.
#define FLASH_PAGE 251

unsigned long *t1p = ADDRESS_OF_PAGE(FLASH_PAGE);

void setup()
{
  // Our buttons need to be inputs, with pullups.
  pinMode(BUTTON1, INPUT_PULLUP);
  pinMode(BUTTON2, INPUT_PULLUP);
  pinMode(BUTTON3, INPUT_PULLUP);
  pinMode(BUTTON4, INPUT_PULLUP);
```

```
  // This is the output pin for the LEDs. I didn't
  //  make a #define for it because this is the only
  //  place we use it.
  pinMode(3, OUTPUT);
  // This function is defined farther down in the
  //  code. It handles writing the current values
  //  of LED brightness to the four LEDs.
  updateLEDs();
  // Serial is only used during debugging.
  Serial.begin(115200);

  // SFM.deviceName and SFM.advertisementData must,
  //  together, contain fewer than 16 characters.
  //  These strings put us at 15. Whew!
  SFM.deviceName = "Bracer of Time";
  SFM.advertisementData = " ";
  SFM.begin();

  // If we have values in the flash memory, we want to
  //  extract them and use those as our defaults. If we
  //  don't have values in flash, the flash value will
  //  read as -1, so we can check that to see whether we
  //  want to use the value in flash or not.
  if ((int)*t1p > 0) t1Default = (int)*t1p;
  timer1 = t1Default;
}

void loop()
{
  // SFM.process() handles the UI processing, if a phone
  //  is connected to the device.
  SFM.process();

  // Timer 1 section
  // Start the timer when the button is pressed and the timer
  //  isn't running.
  if ((digitalRead(BUTTON1) == LOW) && (t1active == false))
  {
    delay(25);          // Debounce the input.
    t1active = true;    // Start the timer.
    t1last = millis();  // Start counting from now.
    timer1 = t1Default; // Use the current default value for
                        //  the timer.
    while (digitalRead(BUTTON1) == LOW)
    {/*Wait for the button to be released*/}
  }

  // Stop the timer if it's running.
  if ((digitalRead(BUTTON1) == LOW) && t1active)
```

```
{
  delay(25);            // Debounce the input.
  t1active = false;    // Stop the timer.
  while (digitalRead(BUTTON1) == LOW)
  {/*Wait for the button to be released*/}
}
// Activates every 1000ms while the timer is running to
//   keep the time updating.
if ((millis() - t1last > 1000) && t1active)
{
  t1last = millis();
  timer1--;
  Serial.println(timer1);
  if (timer1 == 0)
  {
    timer1 = t1Default;
    t1active = false;
    t1LEDVal = 255;
    updateLEDs();
    Serial.println("Timer 1 expired!");
  }
}

// LED blinking section
// Updates 10 times a second, to update the LED of any
//   timer that is running.
if (millis() - LEDUpdateLast > 100)
{
  // First, take a note on the current time, so we
  //   know when to next enter this subsection of code.
  LEDUpdateLast = millis();

  // Update the values being displayed on the LEDs.
  updateLEDs();

  // Now, calculate the values that we'll display on
  //   the LEDs next time through the loop.
  // T1 LED section
  if (t1active)
  {
    // Adjust the LED value for this LED by changing
    //   the index we use from the LEDSteps array.
    t1LEDVal = LEDSteps[t1LEDIndex+=t1LEDDir];
    // "Bounce" the direction of adjustment when we
    //   reach one end or the other of the array.
    if (t1LEDIndex == 0)
    {
      t1LEDDir = 1;
    }
```

```
        else if (t1LEDIndex == 15)
        {
          t1LEDDir = -1;
        }
      }

  }
}

// UI Element object handles
// We could have put this stuff up at the top, but I
//   wanted it closer to the UI function.
uint8_t t1Input;

uint8_t getValuesButton;
uint8_t updateButton;

// This function is a Simblee library function that
//   defines the UI elements that we'll see on the phone.
void ui()
{
  SFM.beginScreen();
  // We need to refetch these values every time we
  //   reconnect to the phone, in case they changed.
  if ((int)*t1p > 0) t1Default = (int)*t1p;
  timer1 = t1Default;

  // These are the text boxes that display the name of
  //   the timer the text field will be controlling.
  SFM.drawText(40,80, "Timer 1:");

  // These are the text fields that allow the user to input
  //   a number to be used for the default value of each timer.
  int temp = -1;
  t1Input = SFM.drawTextField(100, 70, 50, temp);

  // Define two buttons: one to get the values from flash and
  //   populate the text fields, and one to store the values to
  //   flash.
  getValuesButton = SFM.drawButton(40,240,150, "Get settings");
  updateButton = SFM.drawButton(40, 300, 150, "Store settings");
  SFM.endScreen();
}

// This is a Simblee library function that handles events caused
//   by objects in the UI. We have two types of events in this
//   application: text field entry events and button events.
void ui_event(event_t &event)
{
```

```
// First, handle the text field entry events. These occur when
//  the "enter" key is pressed while the cursor is in a text
//  field.
if (event.id == t1Input)
{
  t1Default = event.value;
  Serial.println(event.value);
}

// Now, the update stored values button. This records
//  the values in the fields to flash memory so they
//  persist through power loss or reset.
if (event.id == updateButton)
{
  int rc = flashPageErase(FLASH_PAGE);
  Serial.println(rc);
  rc = flashWrite(t1p, (unsigned long)t1Default);
  Serial.println(*t1p);
  timer1 = t1Default;
}

// This button fetches the current values and puts them
//  into the text fields.
if (event.id == getValuesButton)
{
  SFM.updateValue(t1Input, timer1);
}
}

// Put the current intensity on each LED.
void updateLEDs()
{
  RGB_Show(t1LEDVal,0,0,3);
}
```

USE IT

You'll need to download the Simblee *For Mobile* app to your smartphone to change the timing settings. This app acts like a browser for
Simblee-based projects. It's available for both iPhone and Android
and is free to download.

Using the app is pretty simple: when your bracer is powered
up and your Simblee has been programmed with the project code,
open the For Mobile app and you should see a list item titled "Bracer

of Time." Touch that list item to bring up the interface for the bracer, shown in Figure 9-9, where you can enter the desired delay in seconds for each timer, as well as store the values in flash memory on the bracer so they persist after power down.

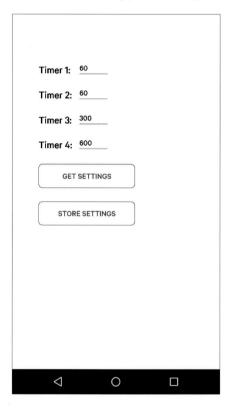

FIGURE 9-9:
The bracer interface

SUMMARY

You should now have a complete magical timing bracer! I hope you find some need for it.

I recommend looking through the full code and the comments left there. You'll find some useful tricks for linearizing the output of an LED, so that it appears to fade smoothly and evenly from all the way off to all the way on, as well as some goodies about executing code in a loop without a lot of busy waiting. You can find a lot more information on the GitHub page for this project at *http://www.github.com/mhord/bracer_of_time/*. All the resources, including the full code, are available at *https://nostarch.com/LEDHandbook/*.

10

WEARABLE LED TEXT-SCROLLING SASH

BY KRISTINA DURIVAGE

IN THIS CHAPTER, YOU'LL MAKE AN LED SASH WITH INTERACTIVE SCROLLING TEXT.

Small and bright LEDs are really fun for making things you can wear! Wearing LED clothes at night is a sure way to stand out. This chapter will show you how to construct an LED sash that scrolls text and lets you interact with it wirelessly. As the final project in this book, it's a little more difficult than the other projects, in terms of both electronics and construction of the sash itself. But the payoff is a sash whose scrolling text you can alter and update on the move through your phone.

GET THE PARTS

This parts list is pretty substantial and many parts have options, so feel free to browse ahead in the chapter for more specifications. With these items, you can put wireless LEDs pretty much anywhere!

NOTE

Supplies and tools for sewing the sash are listed in "Gathering the Materials" on page 208.

Components

- Battery (I'm using a 7.2 V NiMH battery with a capacity of 5,000 mAh. See "Calculating the Battery Capacity You Need" on page 186 to figure out how much current your project will require.)
- Step-down, adjustable voltage regulator
- 8 × 32 NeoPixel RGB LED matrix (Adafruit P/N 2294, SparkFun P/N 13304, or search for "WS2812" on eBay—NeoPixel is a nickname for the WS2812 type of LED)
- Particle Photon board, with or without headers (available from *https://store.particle.io/products/photon*; see "Connecting the Particle Photon Board" on page 196 for more details)
- T-style Deans plug connectors (1 female, 1 male)
- 22-gauge solid hookup wire (black, red, and another color)

NOTE

If you want to make a smaller garment, feel free to use a smaller NeoPixel array. Also, if you want your LEDs to be more flexible, say in the hem of a skirt or a pant leg, I'd suggest using stranded wire rather than solid core.

I'll be showing how to adjust the voltage regulator in this chapter, but you may want to review the beginning of the section "Attaching a Voltage Regulator" on page 190 if you're unfamiliar with these—there are a few different options and it can get tricky.

Tools

Here is a list of tools you'll need for the electronics part of the build, as well as some that are simply recommended to make things easier:

- Soldering iron
- Solder

- Helping hands (or some other kind of vise)
- Alligator clips
- Pliers
- Multimeter
- Electrical tape
- Heat-shrink tubing

BUILD IT

We're going to build this project in stages. First, we'll build what we need for power. Then we'll hook up the Photon and the LED array, because we can only test those things when they have power to run. Once the electronics are all connected up, we'll make a sash that houses the project comfortably.

The power is the least flashy part, but it will be worth it to have a fully mobile project!

Figure 10-1 shows a diagram of the final circuit. If you get stuck at all, use this diagram as a reference.

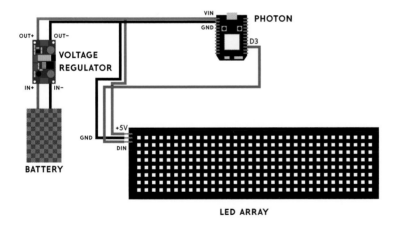

FIGURE 10-1:
Circuit diagram for the text sash

Getting Power: Replacing the Plug on Your Batteries

With our batteries, we'll be dealing with two wires—power and ground. There's one important rule here: *do not let the power and ground wires touch or come in contact with the same piece of metal*. It's a simple rule, but it's very important because if they touch it will cause a short in the wiring, which can mean anything from your project not working to a component actually breaking.

CALCULATING THE BATTERY CAPACITY YOU NEED

Battery capacity is a compromise between weight, cost, and endurance. To figure out what your project needs in terms of battery capacity, first check how many LEDs you have. The array I have is 8 × 32 LEDs, meaning I have 256 LEDs total. Most of the time an LED will use a max draw of 60 mA (full bright white light), with an average use of 20 mA. So if we multiply that by our number of LEDs, our calculations show our sash will have a maximum draw of 15,360 mA (~15.5 A) and an average draw of 5,120 mA (5 A).

However, since our LEDs will display scrolling text, most will be off at any given time, so our actual use will be much less than that. With a multimeter, I measured current draw of between 0.5 and 0.7 A for my final project. On that amp draw, a 3,600 mAh battery will last about 5 hours and a 5,000 mAh battery will last about 7 hours.

If you want to use the array for something else that runs all the LEDs at once and need the higher amp draw, the 3,600 mAh battery will last 13 to 42 minutes and the 5,000 mAh will last 20 to 58 minutes. Those numbers aren't exact—batteries are never exact—and you lose a bit of power with the voltage regulator and the other components, so it's best to overestimate your needs.

For this project, I recommend a NiMH battery that's 7.2 V 5,000 mAh. The LEDs and the Photon need 5 V, and at 7.2 V this battery provides the closest voltage. Later, we'll use a regulator to take the voltage down to 5 V to avoid damaging parts.

Our battery needs to plug into our charger and our project, but chances are they don't currently have the same connection type. We're going to replace the plugs on both battery and charger with T-style Deans plugs so they're compatible. You'll need to be fairly comfortable with soldering for this part.

First, let's label the Deans plugs so we don't mix up the power and the ground. Deans plugs have two terminals, shaped like a T. The horizontal terminal (the top of the T) is usually for the power (red) wire, and the vertical terminal is for the ground (black) wire. Label the Deans plug with a **+** for the horizontal and **–** for the vertical, as shown in Figure 10-2.

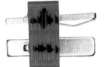

FIGURE 10-2:
Labeled Deans plugs, male
(left) and female (right)

Assuming your battery doesn't come with a Deans plug connector, you'll need to replace its plug with a Deans plug. Again, it's safest
to do this one wire at a time, and put electrical tape around the wire
you're not dealing with.

WARNING
*Don't let the power and
ground wires touch or
come in contact with the
same piece of metal. This
will cause a short in the
wiring, which can prevent
your project from working
or break a component.
Work with one wire at a
time and cover the other
with electrical tape.*

1. **Tin the power and ground wires.** Clip each wire as close to the
    plug connector on the battery as possible, strip a section off the
    end roughly the same length as the plug terminal, twist it so it's
    not fraying, and add solder to the length of the exposed wire all
    around (if you need advice on soldering, see the appendix). This
    process, shown in Figure 10-3, is called *tinning* the wires.

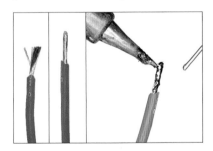

FIGURE 10-3:
Twisting the wire, then
adding solder to it to make
it easier to work with

2. **Prepare the female Deans plug.** Take the female plug (the one
    with exposed terminals coming from only one side), and fit some
    heat-shrink tubing over both the wire and the plug's terminal, as
    shown in Figure 10-4. Set the tubing aside for now; don't slide
    it over the terminal yet. We'll shrink it after attaching the wire to
    the plug.

    If you don't have heat-shrink tubing, you can wrap exposed
    wire in electrical tape. Heat-shrink tubing looks a bit more pol-
    ished, but electrical tape can be removed and reapplied more
    easily.

FIGURE 10-4:
Fitting heat-shrink tubing
(right) on the terminal of the
female Deans plug (left)—
we'll shrink this in a bit

3. **Solder the female Deans plug.** Use a pair of helping hands or a vise to hold the female Deans plug. There is a dent on one side of the terminals from the plug. Heat up one of the terminals of the Deans plug and melt a small pool of solder on the dented side (see Figure 10-5). Be sure to not touch the plastic with the tip of your soldering iron, though! Add solder to the dents on both terminals.

4. **Solder the power wire.** Take the red wire coming out of the battery. Cut two half-inch pieces of the heat-shrink tubing you set aside earlier and slide one piece onto the exposed section of the wire. Push it as far away from the part you'll be soldering as you can, as shown in Figure 10-6—the heat from the soldering iron could start shrinking it before we're ready!

FIGURE 10-6:

The heat-shrink tubing on the wire. We will slide it over the Deans plug terminal and shrink it after attaching the wire to the Deans plug.

All the exposed bits of wire must be covered when we're finished, so make sure the exposed bit of wire is no longer than the terminal—it's better to be too short than too long. Remelt the solder pool on the female Deans plug's positive terminal and push the tinned end of the red battery wire into it until it's covered in solder (Figure 10-7). Remove the iron and hold the wire in place until it hardens, but don't touch it until it's cool.

FIGURE 10-7:
The wire will get hot during this step, so use pliers to hold it close to the Deans plug as you melt the solder and attach the wire.

5. **Solder the ground wire.** Once the joint is cool, repeat this process on the negative terminal with the black battery wire—don't forget to slide heat-shrink tubing onto the wire before connecting it to the Deans plug! We'll shrink the tubes in a moment.

6. **Insulate the wires.** Once both wires are cool to the touch, move the heat-shrink tubing down to cover both the metal terminal and the attached exposed wire. If you don't have heat-shrink tubing that fits, you can cover the components with electrical tape. If you need to try again, you'll need to undo the soldering, put new heat-shrink on, and resolder.

   If you have a heat gun, use that to heat up and shrink the heat-shrink tubing. Otherwise, you can hold a lighter near the tubing to heat it, but be sure not to expose the tubing to a direct flame. Twist the red and black wires to add support to the setup, as shown in Figure 10-8.

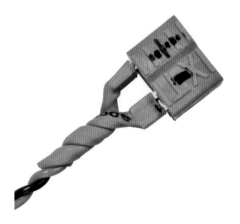

FIGURE 10-8:
The Deans plug with wires

With that, you're done! You've added a different connector onto a battery! Figure 10-8 shows how it should look.

If your charger doesn't currently have a Deans plug, you'll have to repeat this process for the charger but with a male Deans plug for connecting your female battery plug. Make sure to disconnect the charger from power when you put the plug on to avoid any exposed live wires.

Attaching a Voltage Regulator

It's likely that your battery will provide more than 5 V, so we need to reduce that to the 5 V the Photon needs to prevent it from over-powering. Voltage regulators, as seen in Figure 10-9, can step down voltage to the level you need. We'll use an adjustable voltage regulator because they're the easiest to find.

The term *voltage regulator* is pretty intimidating, and the board looks intimidating too, but don't worry! Once you've set it up, it'll do its thing and you'll never have to touch it again.

1. **Buy a voltage regulator.** There are a couple of important things to note when you're voltage regulator shopping. First, buy a *switching* voltage regulator that looks similar to the one in Figure 10-9. Second, make sure to buy a regulator with two blue boxes, not one—this lets you adjust both current and voltage.

    Most voltage regulators return either 3 A or 5 A. Either is fine for displaying text only (if you remember from the section on bat-teries, our draw was 0.5–0.7 A), but if you end up changing the program so most of the LEDs are on, you'll need to be sure that you keep the draw under the maximum the regulator allows.

FIGURE 10-9:

The voltage regulator from the top

2. **Connect the plug.** There are only four parts on this board you need to care about—the two plastic boxes with screws at the top of Figure 10-9, and the IN and OUT boxes on either side.

You'll need four pieces of wire—two black and two red. One set will connect the battery and the voltage regulator, while the other will connect the voltage regulator to the rest of the electronics.

Take one red wire and one black wire and twist them together to make them more manageable, then do the same with the other pair. Strip a quarter-inch or so off all eight ends. Following the same process as we did for the battery, connect one pair of black and red wires to a male Deans plug. The male plugs have metal terminals sticking out of both sides; solder one side onto the side with the smaller metal piece. The female plug you attached to the battery will plug into this, and it will be the incoming side of the voltage regulator.

3. **Check the voltage of the battery.** Get out your multimeter, then take the male Deans plug you just wired to the red and black wires and plug it into the battery's female Deans plug. Be very careful that the two exposed wires on the other end of the voltage regulator never touch each other!

Set your multimeter to measure DC voltage. Touch the red probe to the red wire and the black probe to the black wire and check the reading on your multimeter (see Figure 10-10).

NOTE

The big copper-wrapped wheel is the inductor coil, and it can be awfully loose and wobbly. If yours is loose, run hot glue on either side of it to hold it in place. Hot glue doesn't conduct electricity, so it won't interfere with anything on the board.

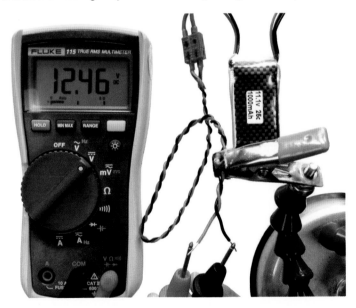

FIGURE 10-10:
Measuring the voltage of the battery, and testing the connections of the wires to the Deans plug

If you aren't getting any reading, try the following:

- Check that your multimeter probes are plugged into the voltage and ground terminals of the multimeter.

- Confirm that your multimeter is set to measure DC voltage.

- Make sure your wires are plugged in to the battery.

- See if you can get a reading from a regular battery, like a AA, by connecting the red probe to the positive end and the black to the negative end, to confirm that your multimeter is working.

- Test the battery more directly: hold the battery's female Deans plug in your helping hands, then put the probes inside the two terminals (with red to positive and black to negative) and see if you get a reading out of that. If not, try pinching the wire where it was soldered to the terminal and see if you get any kind of numbers then—it may be that the solder connection was not firm enough.

- Try pinching the wire where it's connected to the male Deans plug and see if you get a reading then.

- If you still can't get a reading, ask someone to look over your work! Just be sure not to carry this setup around while the battery is plugged in.

Once things are working, unplug the battery, set it aside, and grab the voltage regulator.

4. **Wire up the voltage regulator.** You'll need a Phillips-head screwdriver small enough to fit the screw at the top of the terminals on the IN side, labeled in Figure 10-9. We'll attach the red wire to IN+ and the black wire to IN−.

   Unscrew the screws until the top is flush with the case around it, and place the black wire from your male Deans plug inside IN− to test that it fits (see Figure 10-11). If you can see any exposed wire coming out of the terminal, trim some back—you don't want any bare wire showing, but you want as much wire inside as will fit.

   Once the wire fits, tighten the screw to hold it in place, as shown in Figure 10-11. Repeat this for the red wire in IN+, and then tug on both wires slightly to make sure they won't fall out. You can twist the wires back again to make sure they are close together; just make sure no exposed wire is showing or touching.

FIGURE 10-11:
Screwing in the red wire
after it was cut to fit
so there was no metal
showing

Take the other set of black and red wires not yet attached
to anything and do the same for those in the OUT terminals.
Remember that red goes to OUT+ and black goes to OUT–.
Figure 10-12 shows how it should look.

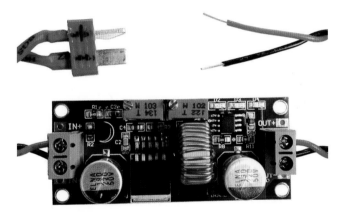

FIGURE 10-12:
Your voltage regulator
so far—male Deans plug
attached to the IN side and
exposed wires attached to
the OUT side

5. **Adjust the voltage to 5 V.** To adjust the voltage, we turn the
screws on the OUT side of the regulator. The left screw adjusts
the output voltage, and the right adjusts the maximum output
current. On some boards the screw terminals will be labeled
CV for volt adjustment and CC for current adjustment (shown in
Figure 10-13). You might want to label them V for volts and A for
amps to avoid mixing them up.

FIGURE 10-13:

CV (left) is for adjusting volts, and CC (right) is for adjusting amps, or current.

Get your multimeter back out, and attach one end of an alligator clip to each of the exposed wires and the other end to each multimeter probe for both positive and negative, shown in Figure 10-14. Again, be very careful to not let red and black touch each other. If you aren't getting any reading at this point, check the voltage regulator connections on both ends.

FIGURE 10-14:

Use alligator clips to connect the end of each wire and the multimeter probe so your hands are free to adjust the voltage regulator.

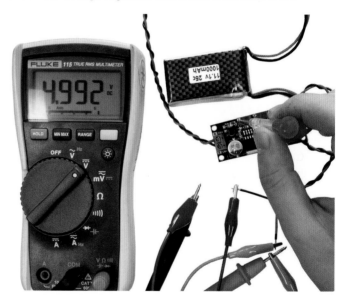

Some voltage regulators have a small light, and if it's lit, it's getting power but not sending it back out. If it's not lit, it's not even getting power, so you should check the battery connection. Be sure you unplug the battery between adjusting wires, though—you don't want to accidentally have ground and power touch somewhere.

Once your multimeter is reading, turn the voltage adjustment screw counterclockwise until the voltage change registers on the multimeter. Continue turning the screw until the multimeter reads 5 V, erring to be a little under 5 V rather than over. If counterclockwise doesn't adjust the voltage in the direction you need, try turning it clockwise and note which direction makes the voltage increase and decrease.

6. **Adjust the current.** Current is measured differently than voltage: voltage needs to be pretty precise, as components like the Photon or the LED array won't work if the voltage gets too far away from 5 V. Current can be set as high as you want, and components will draw what they need. We'll set the current to the max.

   Remember which direction turns the voltage up, and turn the current screw in the same direction. You'll hear a click with every turn when it is maxed out. The good news here is if your current is too low, everything will just shut off without being damaged, and if the current is too high, your parts will draw only what they need. Note which direction you turned this screw for future debugging.

7. **Split the line.** Our LED array takes a lot more power than the Photon can output, so we'll split the line from the voltage regulator so that one set of power/ground wires will power the Photon and one set will power the LED array. We'll accomplish this by forking the wire coming out of the voltage regulator, so the red and black pair becomes two red and black pairs.

   Cut a new piece each of black and red wire, about two feet long. Fold the red wire in half, and strip a small section of wire at the fold. Wrap the exposed metal middle together with the exposed end of the red wire from the voltage regulator and add some solder so they're attached. However you can get those two wires to stay together will work. When you're done, shrink some heat-shrink tubing over that section.

   Repeat this for the black wire. These steps are shown in Figure 10-15.

FIGURE 10-15:

Wrap the exposed end of the red wire around the exposed middle of a longer red wire, solder, and heat-shrink.

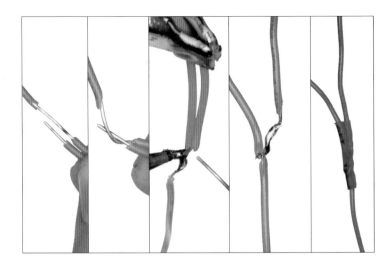

8. **Create two pairs of wires.** Pair up a red and black wire coming out of the fork by twisting them together so you have two pairs, and use more heat-shrink tubing to hold them together at the fork, as in Figure 10-15. Remember to never let exposed black and red wires touch when the battery is plugged in!

With that, we're done with the power part of this project! Figure 10-16 shows a diagram of the project so far. You now can plug in a battery and, once we attach it to the board and LED array, you'll be able to power both components and run them completely wirelessly.

FIGURE 10-16:

A diagram of the components so far

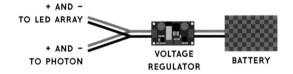

Connecting the Particle Photon Board

This section will just cover the wiring for the Photon; you'll hook the Photon up to your phone as a wireless hotspot and program it in "Code It" on page 200. If you want to get a feel for the Photon before you start semipermanently soldering wires to it, feel free to head there or look at other example projects using the board first. We'll be using just three pins on the Photon—VIN, GND, and D3.

Some Photon boards will come with headers, and some won't—Figure 10-17 shows the difference in how they'll look.

FIGURE 10-17:
A Photon without headers (left) and with headers (right)

Headers allow you to connect the Photon on a more temporary basis by plugging and unplugging it from a breadboard rather than soldering wires. But headers also add more bulk, and it's easier for components to unplug from a breadboard if they get tugged the wrong way.

Without headers, you will solder wire directly to the board. This is more permanent—you can remove the wire by heating up the solder connection again, but that risks damaging the board.

If you want to experiment and are worried about getting things wrong, I recommend using headers and a breadboard, especially if you might want to use the Photon for other things.

If you want to make this project permanent, I'd recommend soldering directly to the board at some point, even if you start with the breadboard. The following instructions assume you're soldering wires directly to the Photon.

1. **Solder the wires.** Take one pair of the forked red and black wires from the voltage regulator and connect them to the Photon, with the red wire to VIN and the black wire to GND, and then solder them in place, as shown in Figure 10-18. Make sure the solder joints from the VIN and GND do *not* touch each other or anything else on the board. Clip any extra wire you have poking up through the board so there's nothing to snag.

FIGURE 10-18:
Soldering the wires directly to the board

2. **Test the plug.** Double-check that the other pair of wires from the voltage regulator is protected (if you stripped the ends, make sure they aren't touching), then plug in your battery. You should see your Photon spring to life! If you're nervous about this step, try plugging the battery in for just a second and make sure you don't see any smoke or sparks. If you *do* see smoke or sparks, somewhere your VIN/red line is touching your GND/black line. Double-check your soldering to make sure everything is connected up correctly and the solder is where it's supposed to be and nowhere else.

Wiring the LED Array

We'll finish up by soldering the second pair of forked wires to the LED array. The LED array will need three lines: power, ground, and the data line from a digital pin (D3).

There are many different brands of LED arrays out there, and the connections may be labeled differently. You'll want to look either for arrows pointing toward or away from the board, or the words IN and OUT, to indicate which side the data line should go on.

There should be three terminals labeled + or 5V, – or GND, and DI (or DIN), as shown in Figure 10-19. If they're not labeled, look at the wires that come attached: red is usually +, black is –, and data is some other color. We'll be removing the wires, so label the board itself if you need to.

1. **Remove the wires the board comes with.** The wires are just soldered on the surface, so put your hot soldering iron on the terminal and lightly pull the wire until the solder melts enough for the wire to come off. Remember not to get solder anywhere but the terminal, and don't touch any other part of the board with your soldering iron.

2. **Connect your power wires.** Next, take the remaining pair of forked red and black wires and connect them to a set of power + and – terminals on the LED array (they may also be labeled 5V and GND). Add a bit of solder to each pad first, then strip a small section of wire, cut it to fit the pad, heat up the solder on the pad, and put the stripped wire under the solder. This step is shown in Figure 10-19.

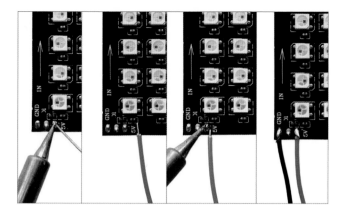

FIGURE 10-19:
Add more solder to the terminals. Make sure the exposed bit of wire is about the same length as the pad, then melt the solder on the terminal to attach the wire to the LED array.

3. **Connect the data line.** Take a wire that's not red or black and connect it from the DIN terminal of the LED array to D3 (or any other D pin) of the Photon, as shown in Figure 10-20. Wire length matters here—the shorter the wire between the Photon and the LED, the less chance of interference, which can mess up the LED display. Solder this connection in the same way as the power connections.

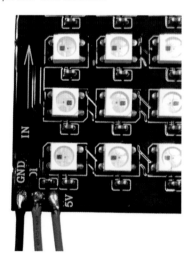

FIGURE 10-20:
Solder the DIN connection.

Test the wiring now by plugging in the battery. You should see the Photon's light turn on. The LEDs won't light up yet because there's no program telling them to. You'll plug the battery in to turn the device on and off from now on.

If your Photon's light does not turn on, check the wiring coming out of the battery and into the Photon. Somewhere a wire is probably broken, meaning the Photon is not receiving power. Use the multimeter to see the voltage level at different places to diagnose where the break occurs. Also check that your red and black wires are not reversed.

If it still isn't working, you may have accidentally turned the current adjustment screw all the way down instead of up. Try turning the current screw all the way in the opposite direction and see if that helps.

Once the Photon powers up, you can put your soldering iron away, because except for adjustments you might make after you've sewn your sash, you're done with the physical hardware part of this project! Next up we'll be uploading some scrolling text code, testing it, and then assembling the garment for the electronics to go inside!

CODE IT

Our end goal with the coding is to be able to text a message to a number and have that message display on the sash. To get this to work, we need code running in two places: on the Photon, and on a server that will wait to receive a message and then send it to your sash. We'll focus on the code running on the Photon first.

You can download the full code for this project from *https://nostarch.com/LEDHandbook/*. I'm going to give an overview of aspects of this code, but I've also added comments in the code itself to document how it works.

Connecting the Photon to Wi-Fi

First, you'll need to get your Photon hooked up to Wi-Fi. Use the documentation from the Photon site at *https://docs.particle.io/guide/getting-started/intro/photon/* to see how to get started. Note that the documentation may be updated between this writing and your reading, so it may look a little different but should be easy to follow.

This step is optional and can be reconfigured later, so if you just want your Photon to connect to the internet right away, connect it to a traditional Wi-Fi network and skip this part. Once your Photon is connected, you should rarely have problems.

You'll need to download the Particle app to connect your Photon to a Wi-Fi network.

1. **Download the app.** Go to your app provider and search for Particle. The tile for the app looks like Figure 10-21. Download and install the Particle app and create your own account.

FIGURE 10-21:

The Particle app tile

You should be taken to the Your Devices screen. Click the **+** in the corner and choose **Set up a Photon**. You should now be able to connect to your Photon.

2. **(Recommended) Connect your Photon to your phone's hotspot.** I recommend turning on your phone's Wi-Fi hotspot ability and connecting your Photon to that. This can be tricky, but it allows you to be fully mobile and not have to rely on hooking up your Photon to multiple Wi-Fi networks.

To connect your Photon to your phone, note your phone's network name and password, then click **Connect to Wi-Fi network** and select **My network is not listed**. Enter your Wi-Fi name and password, as shown in Figure 10-22.

When the app shows a spinning indicator and says it's looking for the Wi-Fi network, go back to Settings and turn your phone's hotspot on. The Photon will connect to the internet, and from that point forward it will connect to your phone's hotspot as long as the hotspot ability is on.

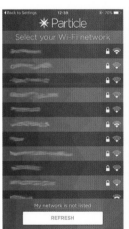

Coding the Photon

Once your Photon is connected, head over to *https://build.particle.io/build*. This is where you'll upload your code to your Photon. You'll set up an account, and then you'll be taken to an editor where you can send messages to your sash!

Adding the LED Library

The code for your Photon is available to download at *https://nostarch .com/LEDHandbook/* in the file *textsash1.ino*, but you'll first need tell Particle to use the library for controlling the LEDs.

In the editor, hit the button to create a new app. Give your app a recognizable name, like TextSashPhoton. Hit the save button (folder icon) on the left side.

The Libraries icon looks like a bookmark on the left side of the screen; click this and scroll to the list of Community Libraries. You'll likely see the NeoPixel library at the top of the list (Figure 10-23).

FIGURE 10-23:

The Libraries icon and the Community Libraries section

Click the NeoPixel library and you should see some code come up. Ignore this for now and look for the Include In App button, as shown in Figure 10-24. Click it and select your recently made app by its name (if it's not there, you might have forgotten to hit save after entering a name).

FIGURE 10-24:

Including the NeoPixel library

You should see an Included Libraries section in the gray bar to the left with NeoPixel listed, as well as an #include line at the top of your file specifying the `neopixel` library. Now you're ready to add more code!

Testing the Photon

Before we add the code that will let us send messages to the array, let's write some code to test that all the wires are hooked up right and everything is working as planned.

1. From the code files downloaded from *https://nostarch.com/ LEDHandbook/* open the file *testArray.ino* and copy it onto the main part of the screen. Be sure the line that includes `#include` is there.

2. Now make sure the Photon app knows which device it's uploading code to. Click the Devices icon, which looks like a scope or crosshairs (see Figure 10-25). You should see your newly paired Photon on the list. Give it a name and then click the name. A star should show to the left of the name—when you upload code now, known as *flashing* code, it'll go to that device.

FIGURE 10-25:

The Devices icon

3. Make sure your Photon is powered and connected to the internet—the LED on the Photon should be blinking cyan if it is—and click the flash button (which looks like a lightning bolt) on the top of the left bar. You should see the Photon blink magenta. If this is your first time uploading code, it might go through some different colors while it makes sure the firmware is up to date and resets. When it's all done, one LED in the corner of the array should light red.

   If the light didn't come on, do some more troubleshooting.

Coding Text to Your LED Display

Once you have the Photon working, paste the code from *shortstr _textsash.ino* from the book's resources into the Photon editor. I've fully documented every function and any confusing bits within the code itself, so feel free to peruse it. Just scroll down past the code at the top that encodes the characters.

Flash that code by pressing the lightning button, and you should see `Hello World!` display across your LED array. That's cool, but now let's make it say exactly what we want!

Please note there is one limitation for now: you can only send the sash 63 characters at a time. Sending more doesn't return an obvious error; it just doesn't work. I'll show you how to get around this later.

To send a message to your LEDs, you'll need two things: your Photon's device ID and your access token. To find the device ID, click the Devices icon on the sidebar of the editor, then hit the arrow to the right of the device name. This will drop down the device ID in a textbox. Copy this ID to a text file for now.

To find the access token, click **Settings** on the bottom left and you'll find the code in a box there. Copy this to a text file too.

We will use these two identifiers to send data to the sash. Just like how your computer gets and sends data from the internet, connecting your Photon to the internet allows you to send data to your specific device to execute code. We'll use *Postman*, an app for Chrome. You can download it and find out more at *http://www .getpostman.com/*.

NOTE

If you have a program of choice to send REST calls, feel free to use it.

Download the Chrome app and open Postman. Now we'll set up a call that will send text to your sash.

1. Select **POST** from the drop-down menu shown in Figure 10-26, and enter **https://api.particle.io/v1/devices/*deviceID*/ buildString** for the URL, with your device ID replacing *deviceID*.

2. Click the Body tab from the menu beneath the URL box, shown in Figure 10-26, and select the **x-www-form-urlencoded** radio button. Enter **access_token** as the key and enter the access token you copied from your Photon account as the value. Then, on a new line, enter **args** as the key and type in the text you want to display on the LED array as the value. Figure 10-26 shows what this looks like with all the parameters set up correctly. Remember that the message is limited to 63 characters for now.

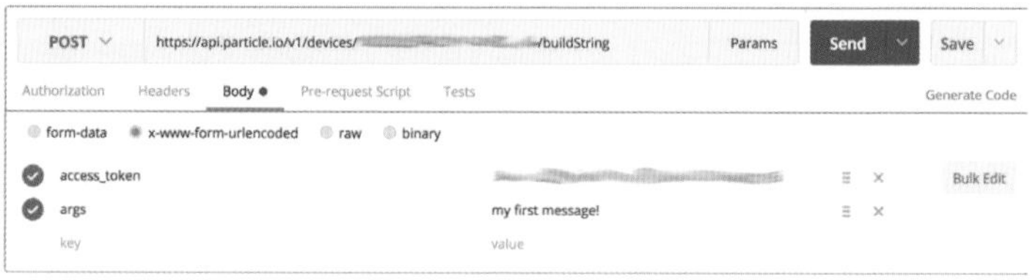

FIGURE 10-26:

How the call to Photon should be set up on Postman

When you're done, save your post and hit the big blue **Send** button. You should see your LED array display your text!

If you don't see anything turn on at this point and everything prior to this has worked, it might be that your voltage regulator's current regulator is set too low. Turn back to "Attaching a Voltage Regulator" on page 190 for a refresher on setting the current on voltage regulators and make the adjustment.

Upping the Character Limit

Now you have an LED array you can communicate with, but you'll want to send more than 63 characters. Go back to your Particle editor and swap out the program you've been using with the file *longstr-textsash.ino*. This will let you make longer strings, but it's a bit more complicated than that.

Instead of sending one message like you did before, you'll send several after each other, and the Photon will store and merge them together until you tell it that you're done and then display the messages.

In a text editor, type what you want to display, but break it up into 60-character chunks. Then, instead of sending one message, send each line as a separate message by pasting it into the args value in Postman. Your first line must be 1,BEGIN and your final line will be 1,END to signal the end of your message. Each line of text needs to begin with 0, and then a space. For example, for the message "This is a really long string that goes on and on and on and I just can't stop myself from typing so much. It's so much fun!" you would send five args via Postman.

The first message tells the sash that it's going to start getting text. The 1,END command tells the sash that you're done sending text and it can display what it's been merging together.

```
1,BEGIN
0, This is a really long string that goes on and on and on and
0, I just can't stop myself from typing so much. It's so much
0, fun!
1,END
```

This might seem a pain to do manually, but don't worry—you're not going to use manually sent messages as a long-term plan anyway.

Coding the Server to Send Messages Through Texts

If you don't mind sending messages to your sash this way, you can stop here. Otherwise, this section will show you how to set up a service that will let you text messages to your sash. This requires a

Twilio account, which is free. Twilio will give you a unique number to send messages to your sash from—if you send them from your own number, you'll end up displaying every text you send! If you get lost at all in this section, you can visit my GitHub repo for directions on where to save information in various files here in the server folder: *https://github.com/gelicia/textSashChapter*.

Installing node and npm

NOTE

You'll need to be running the command terminal as administrator here.

This code will be written in JavaScript, and you'll need to install node and npm. The process might be different depending on what operating system you're using, so follow the instructions at *https://docs.npmjs.com/getting-started/installing-node* to install node and npm for your OS of choice.

1. Once npm is installed, open up a terminal (use the Terminal application in macOS or search **cmd** from Start in Windows and click Command Prompt).

2. Find where node is stored on your computer and navigate to the server folder in the terminal using the cd command; for example:

```
$ cd User/Program Files/nodejs/node_modules/npm
```

3. Run **npm install** from that folder in the terminal and you should see some text like q@, restler@, and nedb@ show up. This is npm installing downloads from the libraries we'll use in our server program, like the neopixel library. You'll also see that the server folder now includes a folder called *node_modules*.

    Inside this *node_modules* folder are some *.js* files. You'll need to alter some permissions here using the device ID and access token from Photon you found earlier.

4. First rename *particleConfigTEMP.js* and *twilioConfigTEMP.js* to **particleConfig.js** and **twilioConfig.js** (i.e., remove *TEMP* from both). The *particleConfig.js* file is easier—you already have the device ID and access token from what you did via Postman.

Setting Up a Unique Number with Twilio

The changes you need to make to the *twilioConfig.js* file will require a Twilio account.

1. Sign up for a Twilio account at *https://www.twilio.com/*, making sure to select that you'll be using SMS (the rest of the questions

don't matter, so you can leave them blank). You'll have to provide a valid cell number to receive a confirmation code, and you'll use that to complete the signup.

2. When you log in, you should see a Get Started button; click that and then click **Get a number**. You should be offered a number to use; click **Choose this number**. Note this number for later. You'll send messages from this number to display on the sash. If you can't find it, scroll down to Phone Numbers from the dashboard and click it.

3. You should see your number listed. Write down your number, then click the home button. You should see your account SID and Auth Token on the page. Put those into the *twilioConfig.js* file you renamed at the beginning of this section—remember you can see the files in my GitHub repo if you need more guidance.

4. Plug your LED array into power and then in your terminal, run **node bot.js**. Now, from your phone, text a message you want to display to your Twilio number.

    You should see the message queue up in the terminal, and then it should show up on the LED array. If something doesn't work, check what error messages you get in your console and confirm that you set up your permissions IDs correctly.

Now you have a simple way to interact with your LEDs! You can either keep your number to yourself and text new messages, or give it to some trusted friends to text as well.

GOING FURTHER WITH THE SERVER

The final step in this section would be to put this program somewhere it won't turn off when your computer shuts down. This is beyond the scope of this book, but if you have the know-how or a friend who does, I recommend using DigitalOcean (they have excellent help for beginners) and the npm library forever (*https://github.com/foreverjs/forever*), which lets you run the *bot.js* script continuously and will restart it if it fails.

ASSEMBLE IT

Finally we'll make the garment to hold the LED display. There are two routes you can take here—either build everything from scratch, or buy something and modify it to suit your project. This is why it's important to do the electronics part first: you need to know what you're putting in pockets before you start making them.

I'll be showing you how to make a sash that holds everything—battery, voltage regulator, Photon, and LED array—in three pockets, as shown in Figure 10-27.

FIGURE 10-27:

The sash is constructed with three pockets, and straps to hold the wiring in place.

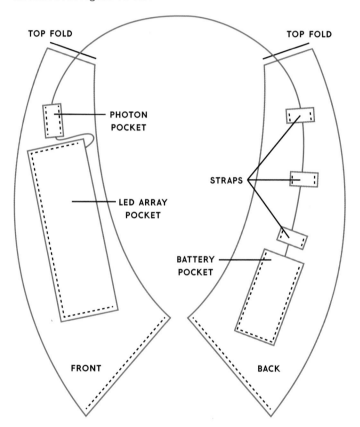

Gathering the Materials

We'll make a black sash, which shows off the colors of the LEDs nicely. However, dark colors can absorb the light and make the displayed text hard to see, so we need two layers—first a thin layer of terry cloth from a cut-up white towel or washcloth, and then the black fabric.

The terry cloth diffuses the light well and counteracts the light absorption from the black fabric. Figures 10-28 through 10-31 show the same LED light display behind different kinds of fabric.

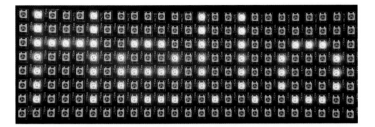

FIGURE 10-28:
The LED array with no cover

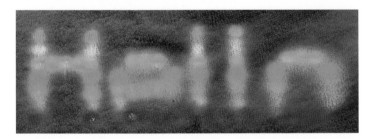

FIGURE 10-29:
The LED array covered by black fabric. Notice the lights are not diffused at all—this is difficult to read when scrolling.

FIGURE 10-30:
The LED array covered by terry cloth. Notice the lights are really diffused and bleed into each other really well.

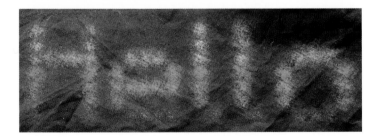

FIGURE 10-31:
The LED array covered by terry cloth and black fabric. This is the final piece.

Supplies

You'll need the following supplies:

- 1.5 yards of black fabric (I used plain black cotton)
- White terry cloth (either bulk or a towel to cut up)
- Newspaper or scrap paper

Tools

You'll also need the following tools:

- Scissors—preferably fabric scissors or pinking shears
- Tape measure
- Marker pen
- Needles, thread, and straight pins
- (Optional) Sewing machine; you could also hand-sew the sash instead
- (Optional) White fabric pen or chalk; this helps with marking where you need to sew
- (Optional) Iron

Constructing the Sash

First, think of where you might put the parts of the project; we need the battery near the base because it's heavy and cumbersome, and if it's at the top it will gradually pull down. We want the LED array front and center, and we need the Photon nearby to minimize interference on the wire. We'll try making the sash out of paper first, to experiment with how things fit.

1. First, measure from the top of your shoulder to your opposite hip to get an estimate of size. For me, 30 inches was about right.

2. Lay two pages of a newspaper or a large sheet of newsprint flat on the ground. The front and back of the sash are the same shape, so you can cut them both out at the same time. Lay the tape measure on the paper in a slight curve, to mimic the shape you'll want your sash (Figure 10-32). Draw that curve on the newspaper, marking the beginning and end points of the length you measured earlier.

3. For the width, think about the width of the LED array, and how much space you want on either side. The LED array is 3.5 inches wide, and I made my sash pattern about 7 to 8 inches wide to begin with. Using the line you drew as the center, mark half your width on both sides (Figure 10-33).

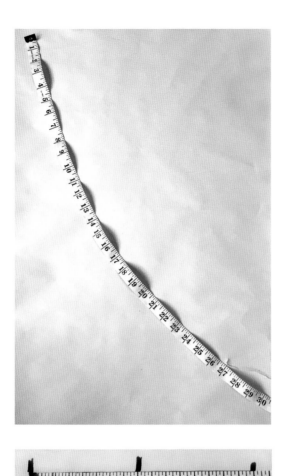

4. You'll end up with two pieces of a vaguely sash-like shape. Cut it
    out of the paper and tape the ends together to form a sash. Put
    this paper sash on (Figure 10-34) and see if you want to make
    any adjustments to the shape. I always take a lot off the shoulder
    and some from the bottom because it sticks out.

FIGURE 10-34:

Try on the paper sash
and trim as needed.

Keep adjusting the shape until you're happy with it. My final
paper sash looked like Figure 10-35.

FIGURE 10-35:

The sash shape once
adjusted

5. Once you have the shape right, you can use it as a pattern to
    cut out the fabric. We'll have two layers to the sash to make it
    stronger, so you'll need to cut out two full sash shapes and sew
    them on top of each other. The best way to do this is to fold your
    fabric with the fold going over the shoulder.

    Mark the outline of the paper on the fabric directly or pin the
    pattern to the fabric and cut around the paper.

    If you have a serger sewing machine, use fabric scissors to
    cut out the sash. If not, use pinking shears to prevent the fabric
    from fraying.

6. If you couldn't fold the fabric over the shoulder, just sew two
    pieces together so it looks like Figure 10-36, with the front and
    back attached at the top. You'll want two of these shapes.

7. Mark the sides as shown in Figure 10-36 and sew the two layers
    on top of each other, either with a serger or using a straight

stitch on the zigzag edges if you cut your fabric with pinking shears.

8. Turn the piece inside out so the sewn edges are now inside. You can iron the pieces flat or stitch on top to make sure they stay flat. You should have something like Figure 10-37 when you're done.

9. Now sew one of the ends shut and tuck the raw edges on the other end inside so it has a nice folded edge. Place the sewn-up end inside the folded end about an inch and stitch both through to close the sash.

Now that the base of the sash is done, we're going to need a few more things: three pockets (one for the battery and the voltage regulator, one for the LED array, and one for the Photon) and straps to hold down the wiring. Let's start with the pocket for the battery and the voltage regulator.

Sewing the Battery and Voltage Regulator Pocket

The battery needs to fit snugly in its pocket so it doesn't move around. To make this work, we'll create a box-like pocket with sides about an inch deep. We'll place this pocket at the bottom of the sash in the back, so it will rest on your lower back. This balances out the weight of the array on the front of the sash. The most important thing here is to make sure the battery fits. I've thrown away plenty of pockets because they were too big or too tight.

1. To make sure the pocket fits the battery and voltage regulator, put the battery in the middle of some fabric and draw around it, leaving a couple inches on all sides and cutting off excess. Fold one side of the fabric over a little to create a hem and sew it in place—that will be the open end of the pocket.

2. On the side opposite the hemmed edge, fold the end and sides in by an inch or so, fold the corners in until the point of the triangle meets the corner of the battery, and then pin the folds in place (see Figure 10-38, top).

3. Fold the sides of the pocket in, so that the triangle made in the corners keeps the sides and bottom up (Figure 10-38, center), and pin it in place. Don't worry if it's ugly; no one will see the end of the pocket.

4. Test the size with the battery, and if it fits well, sew the corner triangles in place, then sew along the outside edge of each corner for good measure (Figure 10-38, bottom).

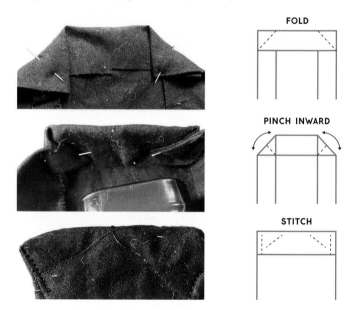

FIGURE 10-38:

Beginning to shape the bottom of the pocket

5. Place the pocket vertically with the open part up. You'll want to place the pocket at the bottom of the back of the sash, since it's heavy and you want it out of the way and out of sight.

6. Pin the pocket in place and stitch the bottom, then the sides, to the sash. Leave a bit of unstitched space at the corners so the case can bulge out a bit if it needs to.

7. Try putting the battery and regulator in the pocket and putting the sash on. You'll have more weight on the other side to even things out later, but if you need to move the battery or use another method to hold it down, feel free to do so.

Sewing the LED Array Pocket

Next up is the LED pocket, which is a flat pocket. We'll add a layer of terry cloth to the inside of this pocket to diffuse the light. So the layers, from your body outward, will be: two layers of black sash fabric, then the LED array, and then the terry cloth and outer black fabric of the pocket.

1. On the terry cloth, mark and cut out around the LED matrix, with a tiny bit of wiggle room, as shown in Figure 10-39. Use pinking shears if you don't have a serger. Leave a bit more room on the top, so you can fold and sew an edge of black fabric over for the top of the pocket.

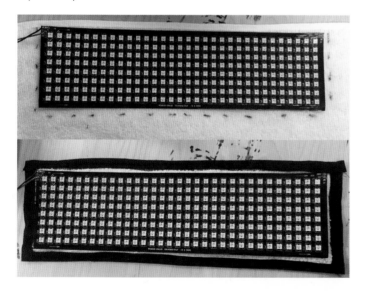

FIGURE 10-39:

Cutting out and sewing the terry cloth and fabric layers

2. Serge the edge of the terry cloth, or leave a zigzag edge if you don't have a serger. We'll be covering the edges in the pocket fabric so it won't be visible. Put the terry cloth on the black fabric and draw around it, leaving 3 to 4 inches around on each side.

3. Fold the black fabric over itself by an inch or so, and then fold it over again on the terry cloth so the serged or raw edge of the terry cloth doesn't show, and pin the folds in place. Do that to all sides, as shown in the bottom panel of Figure 10-39.

4. Place the LED array on the cloth to make sure it still fits.

5. Determine where you want the pocket to go on the sash, line it up, and then pin it in place. This is a good opportunity to try the sash on again, to make sure the pocket looks right when worn and that the array fits. Stitch the pocket in place onto the sash, trying to keep close to the edge of the pocket.

Now we'll build a pocket for the Photon.

Sewing the Photon Pocket

If your Photon doesn't have headers, follow the same idea as the LED pocket to make a flat pocket. If your Photon has headers, follow the same idea as the battery pocket to fit the bulk. You may need to hand-sew the bottom. Then place it on the sash as follows:

1. Once you've made the Photon pocket, position it just above the LED pocket, as shown in Figure 10-40, so that the wire from the Photon to the LED is as short as possible.

FIGURE 10-40:

The Photon pocket positioned a small distance above the LED pocket

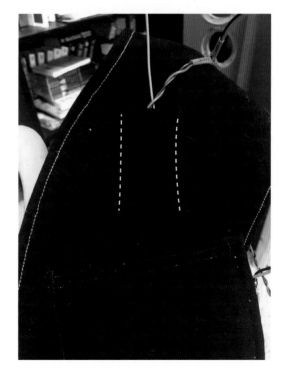

2. Sew only the two long sides, so wires can come out of the top and bottom. They'll hold the Photon in place. If it ever seems like it's poking out too much, you can add a snap to the top of the pocket to keep the Photon in place.

Adding Wire Straps

The last thing we'll add are little straps to hold the wires down to the sash.

1. Cut out a strip of fabric, about 3 inches wide and 1.5 feet long. Serge or sew the long sides together to make a cylinder, then turn it inside out—you might want to attach a safety pin to one side of one end and push it through to the other end to make this easier.

2. Once the right side is out, you'll end up with something more tube-like than strap-like. Stitch the long side opposite the seam or iron it so it stays flat.

3. Cut the fabric into shorter sections, about 3 inches long. You'll probably need about four of them. Hang on to them for now; once you've figured out where your wires will be sticking out in the next section, you'll stitch them to the sash.

4. Now you'll need to make the connecting wires of the project longer or shorter to fit your sash.

    This is your project, so you can make any adjustments you want! I've taken out some seams so wires can come out of the sides of pockets, added snaps to pockets that don't close up, and more.

    To lengthen or shorten the wires, you just need to undo the soldering and either cut down or add new wires, making sure to note where to reconnect them. See the appendix for more soldering tips.

    Remember the voltage regulator doesn't necessarily need to be soldered, so you can unscrew the wires, thread them under straps, and then screw them back up after. Look back at the electronics part of this chapter if you need to remember the wiring. Now you should be able to fit your electronics inside your sash!

5. Once you're done figuring out wire length, space your straps along the sash so the wires lie flat. Pin and then sew the straps in place. I only needed those going from the front over the shoulder

to the back battery pocket. See Figure 10-41 for where I put my straps to hold wires.

See Figure 10-41 for where I put my straps to hold wires.

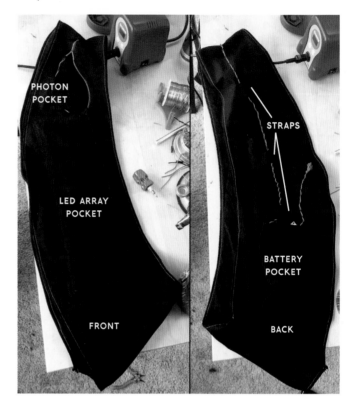

FIGURE 10-41:
The front and back of the sash, with the wires at a good length and positioned under straps. All the components are in their pockets and it's ready to wear!

Once you're done with that, put all the items in their pockets and the wires in their holders, and your sash is finished!

SUMMARY

Now you have a wearable LED array that can display text. I've included a sample server to send text messages, but other options are available too. Small pictures, colors, and designs are all available. Just remember: the more LEDs on at one time, the more current draw. Go out in the world with your own fantastic LED sash!

APPENDIX

GETTING STARTED WITH SOLDERING

BY MATTHEW BECKLER

SOLDERING IS A SIMPLE SKILL, BUT IF YOU'VE NEVER DONE IT BEFORE YOU'LL NEED SOME GUIDANCE. THIS APPENDIX PROVIDES A QUICK INTRODUCTION TO SOLDERING COMPONENTS TO A CIRCUIT BOARD.

Soldering is the process of joining two pieces of metal by melting *solder*, a metal alloy usually made of tin and lead. The solder liquefies, surrounds the two pieces of metal, and solidifies again, forming a strong electrical and mechanical connection. This is a popular method of making connections between electrical components, wires, and circuit boards.

SOLDERING TOOLS

You'll need the following tools for soldering:

- Soldering iron
- Solder (For electronics, not plumbing. Use either lead-free or leaded.)
- Something to solder to, such as a printed circuit board
- Wire cutters
- Parts to solder (These would be wires or electrical components. If this is your first time, I'd recommend soldering wires since they are cheaper.)

You'll need a clear workspace and somewhere heat-proof to rest the iron when you're not using it—you can get soldering iron stands for this purpose. It can also be useful to get a set of Helping Hands to hold your pieces in place while you solder. Beyond these tools, I recommend keeping a wet sponge nearby to periodically clean the tip of the iron.

HOW TO SOLDER IN SIX STEPS

On the printed circuit board (PCB) are *pads*, which are little bits of exposed metal, usually around a hole. This hole is where you'd insert a lead or wire from your component to solder in place.

Before soldering, gather your parts and PCB and position them on a flat surface so they're steady. We'll use a resistor as our example. Follow these steps, using Figure A-1 as a visual guide.

1. Insert the lead from your component into the hole in the pad. Use the tip of your soldering iron to heat *both* the pad and the lead sticking through.
2. Wait a few seconds for everything to heat up.
3. Add solder to the hot joint using a poking motion.

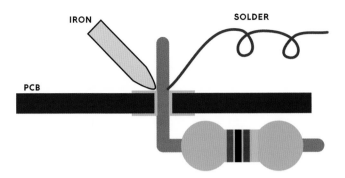

4. Stop adding solder when you have a nice mound of solder around the lead that covers the pad without touching other pads—something like Figure A-2.

5. Leave the iron on the joint for a couple of seconds after you stop adding solder. This ensures that the solder completely flows into the joint. Then remove the iron and let the solder cool.

6. Once the solder is completely cool, trim the excess lead just above the solder and inspect your work.

After soldering each joint, ensure that it is shiny and smooth, as shown in Figure A-2.

FIGURE A-2:
A joint with just the right
amount of solder

There shouldn't be too much or too little solder. It should look like a little volcano of solder around the wire.

KEEPING THE IRON'S TIP CLEAN

After you solder each connection, be sure to keep your iron's tip clean. This improves heat transfer into the connection and makes soldering much easier. Figure A-3 shows an example of a relatively clean solder tip.

When the iron is hot, periodically wipe off the tip with a wet sponge or wet paper towel, and then reapply a little solder to make it shine!

ADDITIONAL RESOURCES

To learn more about soldering, check out the *Soldering Is Easy* comic book by Mitch Altman, Andie Nordgren, and Jeff Keyzer. You can download it for free in the book's resources at *https://nostarch .com/LEDHandbook/* or at *http://mightyohm.com/blog/2011/04/ soldering-is-easy-comic-book/.*

10 LED Projects for Geeks is set in Helvetica Neue, Montserrat, True North, and TheSansMono Condensed. The book was printed and bound by Versa Printing in East Peoria, Illinois. The paper is 70# Anthem Plus matte.

RESOURCES

Visit *https://nostarch.com/LEDHandbook/* for project templates and code files, updates, errata, and other information.

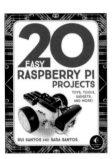

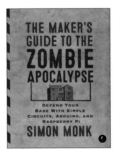